Synthesizers and Subtractive Synthesis 1

Synthesizers and Subtractive Synthesis 1

Theory and Overview

Jean-Michel Réveillac

WILEY

First published 2024 in Great Britain and the United States by ISTE Ltd and John Wiley & Sons, Inc.

Apart from any fair dealing for the purposes of research or private study, or criticism or review, as permitted under the Copyright, Designs and Patents Act 1988, this publication may only be reproduced, stored or transmitted, in any form or by any means, with the prior permission in writing of the publishers, or in the case of reprographic reproduction in accordance with the terms and licenses issued by the CLA. Enquiries concerning reproduction outside these terms should be sent to the publishers at the undermentioned address:

ISTE Ltd
27-37 St George's Road
London SW19 4EU
UK

www.iste.co.uk

John Wiley & Sons, Inc.
111 River Street
Hoboken, NJ 07030
USA

www.wiley.com

© ISTE Ltd 2024

The rights of Jean-Michel Réveillac to be identified as the author of this work have been asserted by him in accordance with the Copyright, Designs and Patents Act 1988.

Any opinions, findings, and conclusions or recommendations expressed in this material are those of the author(s), contributor(s) or editor(s) and do not necessarily reflect the views of ISTE Group.

Library of Congress Control Number: 2023946436

British Library Cataloguing-in-Publication Data
A CIP record for this book is available from the British Library
ISBN 978-1-78630-924-2

Contents

Preface xi

Introduction xv

Chapter 1. Sound Synthesis 1

  1.1. The art of creating sound 1
  1.2. Some reminders 2
    1.2.1. Sound: a bit of theory 2
    1.2.2. Intensity 4
    1.2.3. Pitch of a sound 7
    1.2.4. Timbre 9
    1.2.5. The ear 10
  1.3. Sound typology 21
    1.3.1. Sounds and periods 22
    1.3.2. Simple and complex sounds 23
  1.4. Spectral analysis 25
    1.4.1. The sound spectrum 25
    1.4.2. Sonogram and spectrogram 27
  1.5. Waveforms 28
    1.5.1. Sine wave 29
    1.5.2. Square wave 29
    1.5.3. Rectangular wave 31
    1.5.4. Pulse wave 35
    1.5.5. Triangular wave 35
    1.5.6. Sawtooth wave 37

1.6. Timbre. 39
    1.6.1. Transient phenomena 40
    1.6.2. Range. 41
    1.6.3. Mass of musical objects 42
    1.6.4. Classification of sounds 43
1.7. Sound propagation. 44
    1.7.1. Dispersion 44
    1.7.2. Interference 45
    1.7.3. Diffraction 48
    1.7.4. Reflection 50
    1.7.5. Reverberation (reverb). 51
    1.7.6. Absorption 54
    1.7.7. Refraction 55
    1.7.8. Doppler effect 55
    1.7.9. Phase and beat. 57
1.8. Noise. 58
    1.8.1. White noise 59
    1.8.2. Pink noise 59
    1.8.3. Red noise. 60
    1.8.4. Blue noise 61
    1.8.5. Purple noise 62
    1.8.6. Gray noise 62
    1.8.7. Other noise. 63
1.9. History of sound synthesis 63
1.10. Conclusion 70

Chapter 2. The Different Types of Synthesis 71

2.1. Subtractive synthesis 71
2.2. Additive synthesis. 74
2.3. FM synthesis 78
2.4. Digital synthesis, sampling and wavetables 88
2.5. Physical modeling synthesis 90
2.6. Granular synthesis 93
2.7. Amplitude modulation synthesis 96
2.8. Phase distortion synthesis 98
2.9. Other types of sound synthesis 99

Chapter 3. Components, Processing Techniques and Tools 101

3.1. Oscillators. 101
    3.1.1. Voltage-controlled oscillators 102

3.1.2. Digitally controlled oscillators	103
3.1.3. Digital oscillators	104
3.1.4. Low-frequency oscillators	104
3.2. Filters	105
3.2.1. Low-pass filters	106
3.2.2. High-pass filters	108
3.2.3. Band-pass filters	108
3.2.4. Band-stop filters	110
3.2.5. Resonance	110
3.2.6. Other filters	111
3.3. The envelope generator	112
3.3.1. Attack	114
3.3.2. Decay	114
3.3.3. Sustain	115
3.3.4. Release	115
3.3.5. Other parameters	116
3.4. Amplifiers	117
3.5. Sample and hold	118
3.6. Ring modulator	120
3.7. Waveshaping	122
3.8. Special effects	124
3.8.1. Pitchbend	124
3.8.2. Glide	125
3.8.3. Keyboard tracking	126
3.8.4. Reverb and delay	127
3.8.5. Phaser, chorus and flanger	128
3.9. From monody to polyphony	129
3.10. Controllers	132
3.10.1. Modwheel	132
3.10.2. Breath controller	133
3.10.3. Expression switch and pedal	133
3.10.4. Keytar	134
3.10.5. Other controllers	135

Chapter 4. Work Environment — 137

4.1. Materials	137
4.1.1. ARP 2600	138
4.1.2. The Minimoog	147
4.1.3. The Behringer Neutron	152
4.1.4. The Novation Bass Station II	156
4.1.5. The Arturia MatrixBrute	160

4.2. Software.. 167
   4.2.1. Native Instruments Reaktor........................... 167
   4.2.2. VCV Rack 2.. 168
   4.2.3. Cycling '74 Max/MSP................................. 170
   4.2.4. Pure Data... 171
4.3. Conclusion... 173

Chapter 5. CV/Gate and MIDI 175

5.1. CV/Gate.. 175
   5.1.1. Overview.. 175
   5.1.2. Operation... 176
   5.1.3. Note definition 176
   5.1.4. Operation of the gate (or trigger).................. 180
5.2. Musical Instrument Digital Interface....................... 180
   5.2.1. MIDI version 1.0 180
   5.2.2. MIDI Version 2.0.................................... 181
   5.2.3. Principle 183
   5.2.4. The hardware 184
   5.2.5. The software.. 187
   5.2.6. MIDI Control Change 191
   5.2.7. Examples of MIDI transmission 194
   5.2.8. MIDI implementation chart........................... 198
   5.2.9. General MIDI standard 199
   5.2.10. The General MIDI 2 standard........................ 200
   5.2.11. The GS format...................................... 201
   5.2.12. The XG format...................................... 202
   5.2.13. MIDI file structure 203
   5.2.14. An example of a MIDI file.......................... 211
5.3. MIDI CV/Gate converters.................................... 213

Conclusion.. 215

Appendix 1. General MIDI 1 and 2 Instruments...................... 217

Appendix 2. MIDI Box, Merger and Patcher 235

Glossary 237

References 245

Index 255

Preface

If you want to know if this book is for you, how it is constructed and organized, what is in it, and what conventions will be used, you have come to the right place, this is the place to start.

Target audience and prerequisites

This book is intended for all those who are interested in sound synthesis and synthesizers, whether they are amateurs or professionals, or even musicians, performers or composers.

The information presented in some sections requires basic knowledge of general computing and digital audio.

For some work on microcomputers, you will need to have good knowledge of the operating system (paths, folders and directories, files, names, extensions, copies, moves, etc.).

For exercises based on the VCV Rack and Native Instruments Reaktor Blocks software synthesizers, you will need to know their philosophy, general principles of design and use in order to build a VCV or Reaktor Software Modular Rack.

For exercises related to the visual programming languages Max/MSP and Pure Data, basic knowledge of their interface and the commands of their editors will be necessary.

If you do not feel comfortable with these prerequisites, a set of books and tutorials are mentioned in the reference section of this book, which will help develop your knowledge.

Possession of a synthesizer based on subtractive synthesis will be a plus, especially if it is an ARP2600, a Minimoog, a Novation Bass Station II, a Behringer Neutron or an Arturia MatrixBrute. Software or hardware clones of these machines are also welcome.

Software such as Pure Data or VCV Rack can be downloaded easily and for free, as can software clones of some synthesizers (Minimoog, ARP 2600). Consult the links in the reference section of this book for this purpose.

Organization and contents of the book

This book consists of two volumes:

1) *Synthesizers and Subtractive Synthesis 1: Theory and Overview*;

2) *Synthesizers and Subtractive Synthesis 2: Application and Practice*.

Volume 1 successively presents a preface, specifying the contents and the writing conventions used, and then an introduction followed by five chapters, a conclusion and two appendices:

– sound synthesis;

– different types of synthesis;

– components, processing and tools;

– work environment;

– CV/Gate and MIDI.

The conclusion, as its name suggests, attempts to assess the current state of subtractive synthesis and synthesizers.

Appendices 1 and 2 provide some additional elements and some reminders. You will find information in the following order:

– General MIDI 1 and 2 instruments;

– MIDI boxes, mergers and patchers.

Volume 2 presents a preface identical to that of Volume 1, followed by five chapters, a conclusion and four appendices:

– subtractive synthesis, the beginnings;

– subtractive synthesis, the fundamentals;

– advanced subtractive synthesis;

– duophony, paraphony and polyphony;

– sequencer and arpeggiator.

Appendices 1– 4 provide some additional information in the following order:

– USB connectivity;

– Pure Data extensions;

– keyboards and interface;

– MIDI notes, numbers and frequencies.

The conclusion sheds light on the contents of the book and a brief overview of the future evolution of sound synthesis systems and software.

At the end of this book, you will find references and a list of internet links.

A glossary is also present, and it will explain certain acronyms and terminology very specific to sound synthesis and synthesizers.

Each of the chapters can be read separately. If concepts that are dependent on another chapter are present, the references to the relevant sections are indicated. However, Chapter 1, devoted to sound synthesis, provides the necessary foundations for understanding the subsequent chapters.

If you are a new reader on the subject, I strongly advise you to read Chapter 1 first; the following chapters will then be clearer.

For everyone else, I hope you will discover new notions that will enrich your knowledge.

Conventions

This book uses the following typographical conventions:

– *Italics*: reserved for important terms used for the first time in the text, which may be present in the glossary at the end of the book, mathematical terms, comments, equations, expressions or variables.

– UPPER CASE: reserved for command names, entry, exit, or connection points, specific functions, modules belonging to the different hardware or software synthesizers used in the exercises. It can also be elements, options or choices within menus present in the interface of a program.

– `Courier` font: reserved for objects manipulated within the visual programming software Max/MSP and Pure Data.

Notes are indicated by the presence of the keyword:

NOTE.– These notes complete the explanations already provided.

Figures and tables all have a description which is often useful for understanding.

Vocabulary and definitions

As with all techniques, subtractive sound synthesis and synthesizers have their own vocabulary, with words, acronyms, abbreviations, initials and proper nouns not always familiar. This is the role of the glossary already mentioned above.

Acknowledgments

I would particularly like to thank the ISTE team, and my editor Chantal Ménascé, who trusted me.

Finally, I would like to thank my wife, Vanna, and my friends, passionate about the subject, who supported me throughout the writing of this book.

October 2023

Introduction

Since 1974, the year I purchased my first synthesizer, a Roland SH1000, I have always been passionate about sound synthesis.

In 1976, I acquired a Minimoog, a machine that I still own today and with which I learned the basics of subtractive synthesis.

In 1988, during a lunch with members of IRCAM (the French Institute for Research and Coordination in Acoustics/Music), I heard about Max, developed by Miller Puckette and the music school[1] in Chalon-sur-Saône, France, who invited me to a discovery session of this software, then in its version 2.0, not yet marketed.

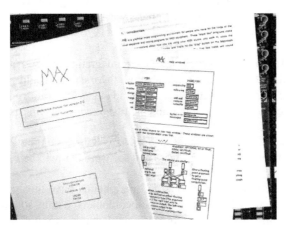

Figure I.1. *Vintage, the Max reference manual version 2.0 (IRCAM 1988)*

---

1 Today the *Conservatoire à rayonnement régional (CRR) du Grand Chalon.*

For me, it was a revelation, as was the same year MIDI-Lisp, a MIDI programming environment for Apple Macintosh.

I invested a lot of time working on Max, MIDI-Lisp and the MIDI standard. It was a great time, the beginning of home studios and software dedicated to music on microcomputers (Atari 520 and 1040 ST, Apple Macintosh 512, Macintosh Plus, etc.).

Year after year, with my home studio growing, I acquired many other machines: synthesizers, drum machines, sequencers, samplers, groove machines, processing, synthesis, programming software, etc.

Over the years, my passion for subtractive synthesis has not diminished and has become part of my professional activity. Even today, it is part of my daily work as a consultant, teacher, trainer and sound designer within my studio.

This book was designed to try to convey my passion and for those interested in the field of electronic music to discover or learn more about subtractive sound synthesis.

Volume 1 brings together theoretical knowledge, and Volume 2 brings together practical exercises on hardware or software synthesizers of several categories, wired, semi-modular or modular.

Although this book is made up of several hundred pages, it is far from covering all the topics related to subtractive synthesis; several thousand pages would not be enough.

In Volume 1, I tried to keep it simple by integrating the scientific notions that seemed essential, without sinking into abstract theories that could put many readers off. This volume details different types of sound synthesis, followed by an overview of the tools and components used to implement subtractive synthesis.

In order to anticipate the practical work covered in Volume 2, a study of the work environment presents the hardware and software that will be covered.

As the links between synthesizers and peripherals have always been complex, the arrival of digital technology and microcomputers within sound synthesis has created a real shift, which has forced manufacturers to define standards and protocols for communications between machines. The CV/Gate control method, dating from the late 1970s, and the MIDI standard, which appeared 10 years later, have become essential and indispensable. It seemed appropriate to specify their principle and define their operation in Chapter 5 of Volume 1.

Figure I.2. *MIDI and CV/Gate Ports (Arturia Keystep). For a color version of this figure, see www.iste.co.uk/reveillac/synthesizers1.zip*

Volume 2 is very practical: it presents 16 exercises performed partly on Behringer 2600 or ARP 2600, Minimoog, Novation Bass Station II, Behringer Neutron and Arturia MatrixBrute for hardware synthesizers. When it comes to software synthesizers, the modular VCV Rack and Native Instruments Reaktor complete the list. A large part of the exercises is reserved for the visual programming environments Max/MSP and Pure Data.

I wanted to put these exercises within the reach of as many people as possible, by choosing some affordable machines (Neutron and Bass Station II) in terms of cost and two open source and free software (VCV Rack and Pure Data).

The 16 exercises are of increasing difficulty and cover, over five chapters, the most important areas related to the practice of subtractive synthesis. Depending on their specificities, some exercises are not required for all synthesizers, hardware, software or languages covered in this book.

Chapters 1 and 2 focus on the key elements of a monodic subtractive synthesizer, oscillator, filter, envelope generator, low frequency generator and noise generator.

Chapter 3 covers more advanced features available only on certain synthesizers: ring modulation, sample and hold and sound effects.

Chapter 4 covers a topic that has long perplexed synthesizer manufacturers because its implementation, until the mid-1980s, required sophisticated and expensive electronics, polyphony.

Chapter 5 is not related to subtractive synthesis, but rather to the tools used in its design and use, sequencers and arpeggiators.

After this brief introduction, it is time to get to the heart of the matter and get to work or, rather, practice my passion!

1

Sound Synthesis

This chapter is devoted to concepts that are essential to the study and understanding of sound synthesis. Here, the reader will find the vocabulary and fundamentals necessary to approach the different processes that produce sound, regardless of the type of synthesis used.

A short history of the different machines that accompanied the birth of sound synthesis concludes the chapter.

1.1. The art of creating sound

Sounds surround us, propagating through air, liquids and solids, but in a vacuum, they do not exist. Each sound has an identity that allows it to be recognized. Our ear, however, is only able to perceive a part of the sound messages, included in a so-called audible frequency range to which I will return in section 1.2.5.

Sound, before being music, is primarily a means of communication for both humans and animals. Throughout time, humans invented tools to create sound using archaic methods, for example tapping on stones or pieces of wood before shaping increasingly sophisticated instruments. Through this evolution a new language was born, which can be referred to as music.

Until the beginning of the 20th century, sounds were always associated with the natural environment, even though musical instruments had become very sophisticated. The invention of the first sound recording methods led to the birth of sounds that could be described as artificial; as for the synthesis part, it was necessary to wait for the development of electronics so that it could take off.

Today, sound synthesis is a real creative tool for composers, musicians and many professionals. It has become an art, even creating new disciplines such as sound design.

In sound synthesis, the objective is not to generate musical notes but to produce a sound, whether that of a real instrument, organ, piano, trumpet, violin, cymbal and many others, or even one that characterizes a steam train, a creaking door, a police siren, a fog horn, a refrigerator, the wind, a storm, the rain, the barking of a dog and even a completely artificial audio phenomenon, non-existent in the natural environment, like an intergalactic battle in a science fiction film[1].

1.2. Some reminders

To begin with, I will describe the nature of a sound, then some of its characteristics, followed by how our ears work. I will continue with an analysis of *sound typology*, *spectral analysis* and *timbre*.

I deliberately chose to present the different mathematical equations in clean and simple forms, without going into detail and avoiding demonstrations. However, the scientific rigor essential in all physical sciences has been retained.

Next, I will approach the fundamental aspects of sound propagation with the presentation of some common phenomena, and then in section 1.8, I will return in detail to the notion of noise.

I will conclude with a history of sound synthesis and the associated instruments.

1.2.1. *Sound: a bit of theory*

What is sound? Answering this simple question is not so easy. We can approach the subject in two ways: the first from a purely scientific point of view, associated with the laws of physics, and the second according to sensory feelings.

For a physicist, a sound is a *mechanical wave* that propagates as a disturbance in an elastic medium or in a body. A wave is considered a back and forth movement (mechanical oscillation) of particles around a resting position, unlike an electromagnetic wave, which propagates energy in the form of an electric field coupled to a magnetic field.

---

1 Which would be very improbable since, in the sidereal vacuum, sound phenomena do not exist.

Figure 1.1. *An example of easily representable mechanical waves. Here, they are created on the surface of the water, after throwing pebbles into it. For a color version of this figure, see www.iste.co.uk/reveillac/synthesizers1.zip*

For most of us, it is much easier to define a sound as an auditory sensation.

Sounds are produced by vibrating objects. These objects are sources, and the environment in which the sound is emitted carries the sound to our ears. At this moment, our brain allows us to perceive them, become aware of them and interpret them.

Most of the objects around us can produce sound when they undergo a shock, friction, blast or deformation. Who has not had fun vibrating a plastic ruler on the edge of a table?

Figure 1.2. *Ruler vibrating on the edge of a table. For a color version of this figure, see www.iste.co.uk/reveillac/synthesizers1.zip*

In a vacuum, a sound cannot propagate because no medium is available to transmit the vibration.

Like all physical phenomena, a sound can be characterized. Many parameters such as *intensity*, *pitch* and *timbre* will be able to define and differentiate it; however, these are not the only ones because the listener can interpret it themselves when listening. This approach implements subjective phenomena called *psychoacoustics*, which are based on the physiology, culture and ethics of the individual who receives the sound message. In this case, the deciphering of the characterization process is still very complex. We are touching on areas where science does not yet have all the answers.

Over the centuries, philosophers have often asked themselves the question: does a sound exist if there is no one to hear it?

The science that physically studies sound phenomena is acoustics. Without being exhaustive, it aims to characterize the *audibility* of a sound, define the means implemented for its transport and transformation or determine the deformations that it can undergo. It is a science that often remains very theoretical. Acoustics is a fundamental basis but when sounds become music, it goes beyond theoretical understanding. Music is not a science; it uses many parameters that are not always measurable, and whose combination is of great complexity.

To continue, I will review certain notions that should not escape those who want to understand the foundations and the nature of sound.

1.2.2. *Intensity*

This parameter characterizes the strength of a sound, determining whether the sound seems loud or quiet. The term loudness is also used.

When a sound is emitted, the sound wave deforms the carrier fluid (air in most cases). This deformation causes a change or a local disturbance in the pressure around the point of emission. This disturbance moves through the surrounding material(s) at a speed (*speed of sound* or *velocity*) that is dependent on the nature of the elements crossed, their state and their thermodynamic properties. The materials traversed, whether fluids or soft or hard bodies, all have a certain *elasticity* so that, in most cases, they regain their initial appearance after having been crossed by the sound. Permanent deformation or destruction may occur if the generated sound pressure is greater than the *elastic limit* of the bodies crossed. This scenario is unlikely, given the low constraints imposed.

The pressure variation mentioned above, mainly observed in the air, is called instantaneous sound pressure. This sound pressure induces sound energy. Both are described by the *acoustic sound pressure level*, also called the *sound level*.

The scale of sound pressure is very extensive, and the unit of measurement is the *pascal* (Pa) (1 pascal = 1 newton/m^2). Normal atmospheric pressure at sea level is defined as being 1.013×10^5 pascals, or 1,013 hPa (1 hPa = 100 pascals). To work on the acoustic sound pressure scale, standardization has retained a reference pressure close to the average absolute intensity threshold of the human auditory system between 1,000 and 4,000 Hz[2], that is, 2×10^{-5} Pa, which can be translated as a power equal to 10^{-12} W³/m². However, these elements remain difficult to handle and are not very demonstrative independently of the fact that the ratio between the minimum audible sounds and very loud sounds is 1/10,000,000.

Sound sensation is above all a physiological phenomenon, and it would be interesting to quantify a sensation of excitation that would consider the extent of the acoustic level of the human ear, which varies according to a logarithmic scale. The inventor of the telephone, Graham Bell[4], defined a base unit, the *bel*, of which a tenth, the *decibel* (dB) is commonly used to quantify sound phenomena in a simple way:

$$N_{dB} = 20\log_{10}\frac{P}{P_0} = 20\log_{10}\frac{P}{10^{-5}}$$

or:

$$N_{dB} = 10\log_{10}\frac{W}{W_0} = 10\log_{10}\frac{W}{1 \times 10^{-12}}$$

where:

– N is the sound level in decibels;

– P is the measured acoustic pressure;

– P_0 is the reference acoustic pressure (0 dB);

– W is the measured power;

– W_0 is the reference power.

The logarithmic progression of the decibel scale results in a doubling of the sound intensity each time it increases by 3 dB.

---

2 Unit of measurement of the frequency of a phenomenon with a period equal to one second, see section 1.1.3.

3 Power measurement unit. A watt is the power received by a system to which an energy of 1 J per second is regularly transferred.

4 Alexander Graham Bell (1847–1922) was an American inventor of British origin (born in Edinburgh), inventor of the telephone in 1876 and founder of the telephone company that bears his name, and creator of the National Geographic Society.

Type of audio signal	Effects	Sound level (dBA)
Launch of a rocket	–	180
Turbojet engine, airplane taking off	–	140
Gunshot, engine test bench	–	130
Formula 1, jackhammer	Pain threshold	120
Rock orchestra, boiler workshop	–	110
Train passage, circular saw, disco	–	100
Walkman at maximum volume, sander, scream	Beginning of risk for hearing risk threshold	90
Radio at maximum volume, machine	–	80
Noisy dining room, busy open office space	Office work	70
Lively conversation, street, public place	–	60
Quiet conversation, large quiet office	Intellectual work warranting great concentration	50
Quiet apartment, quiet office	–	40
Forest walk	–	30
Peaceful countryside, whisper	–	20
Recording studio	–	10
Silence	Audibility threshold	0

Table 1.1. *Table of sound power*

When measuring the sound level using a sound level meter, the unit used is the decibel A dB(A). Indeed, as will be explained in section 1.2.5.2, the human ear has a different sensitivity with respect to the frequency of the sound signals emitted.

To overcome the physiological behavior of the ear, sound power is weighted according to the frequency components of the sound waves captured, before adding them to provide an overall measurement. The dB(A) takes this weighting into account.

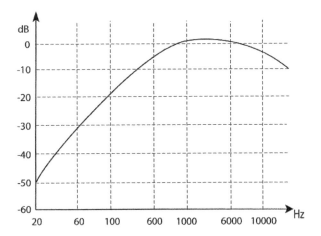

Figure 1.3. *Psophometric chart (weighted curve) dB(A)*

1.2.3. Pitch of a sound

The pitch of a sound is characterized by its *frequency*. It is the number of oscillations per second of the molecules of the body crossed (generally the air) around their position of equilibrium during the passage of a sound wave.

The unit of measurement is the *hertz* (Hz), and its multiples are the kilohertz (kHz), the megahertz (MHz) and so on. The range of frequencies audible to humans extends from 20 to 20,000 Hz (20 kHz). Depending on the individual and their age, this range may vary.

The speed at which a sound wave travels – also called the speed of sound – through air at 20°C is 343 m/s. It varies according to the nature of the body or the medium crossed, the pressure and the temperature. For example, it is 331 m/s in air at 0°C. Through a liquid or solid body, it is greater (~1,400 m/s in water and ~5,000 m/s in steel).

When defining the frequency, we must also consider other parameters such as *wavelength*, *period* and *amplitude*.

Wavelength is the distance between two successive peaks of a sound wave. It separates two successive periods of a periodic wave (see Figure 1.4); it is therefore also the distance traveled during a period.

The period is the time in seconds taken to complete a complete oscillation (a cycle).

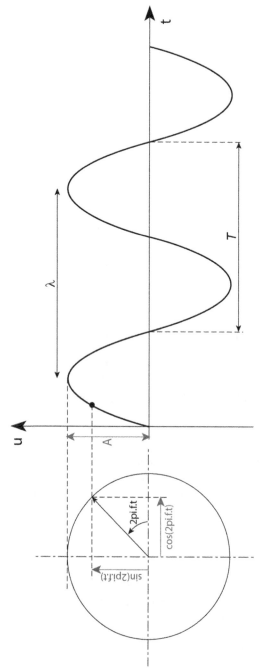

Figure 1.4. *The temporal representation of a sound wave with its different parameters. For a color version of this figure, see www.iste.co.uk/reveillac/synthesizers1.zip*

The period is the opposite of the frequency and vice versa:

$$T = \frac{1}{f} \text{ and } f = \frac{1}{T}$$

where:
- T is the period in seconds;
- f is the frequency in hertz.

For a frequency of 1000 Hz (1 kHz), the period is 0.001 s, that is, 1 ms.

The amplitude defines the sound intensity and, as seen previously, it characterizes a pressure variation.

The wavelength, period, frequency and speed of sound are related by:

$$\lambda = cT = \frac{c}{f}$$

where:
- λ is the wavelength in meters;
- c is the speed of sound in meters/seconds;
- T is the period in seconds;
- f is the frequency in hertz.

1.2.4. Timbre

This is the characteristic that allows our ear to distinguish and recognize a sound, whether it is a common noise, a musical instrument or the voice of a person.

Timbre is largely related to the shape of the sound wave. It is a complex notion that is still difficult to explain today because it implements concepts related to listening, our faculties of appreciation, our sound memory and our level of perception.

To try to describe timbre in more detail, I would have to present many other parameters related to sound such as the physiological functioning of our ear, sound typology, spectrum, transient phenomena, the nature of sound source(s), emission and so on.

Claude Elwood Shannon[5], a notable mathematician, said: "The timbre is what makes a sound sing in our ear".

1.2.5. *The ear*

Among our five senses, hearing is the second most important. It partly relies on our auditory system, whose primary external organ is the ear.

In this section, we will see more precisely how the ear works, its interesting features and we will analyze how the ear works within sound-based environments.

1.2.5.1. *How the ear works*

The ear is divided into three parts: the *outer ear*, the *middle ear* and the *inner ear*.

The outer ear is the sound capture system, it is a system that performs amplification and protection functions. It is separated from the middle ear by a thin flexible membrane, the *eardrum*, which will deform under the effect of sound waves.

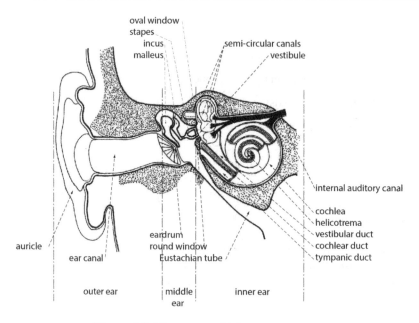

Figure 1.5. *The outer, middle and inner ear*

---

5 Claude Elwood Shannon (1916–2001) was an American engineer, mathematician and cryptographer. Considered the father of digital transmission of information, he is the author of *A Mathematical Theory of Communication*.

It consists of the *auricle* followed by the *auditory canal* with a length close to 2.5 cm. The latter brings sound vibrations to the eardrum and amplifies by 10–15 dB, the average frequencies between 1,500 and 7,000 Hz. These frequencies are the most useful to us, such as those of speech.

The auricle also plays a role related to the positioning of sound sources in space for frequencies between 2,000 and 7,000 Hz.

The middle ear is a cavity filled with air. It is connected to the *pharynx* by the *Eustachian tube*, which opens during swallowing. This has the function of balancing the pressure on both sides of the eardrum. Within the cavity are the *ossicles*, which includes the *malleus* joined to the eardrum, the *incus* and the *stapes*. It is the stapes that acts as an interface between the air medium of the middle ear and the liquid medium of the inner ear. It is attached to the oval window, which separates the inner ear.

The ossicles form a lever that increases the pressure and therefore the amplitude of the sound waves. The surface of the eardrum is approximately 15 times larger than the oval window, which provides an increase of 20 dB. The middle ear is a pressure amplifier.

When a sound is too intense, greater than 80 dB, the stapedius muscle contracts (ossicular reflex) in order to reduce the vibration of the ossicles and thus reduce the transmission of sound waves to the inner ear. This protection ensures a reduction in the sound signal by 40 dB.

It should be noted that the stapedius muscle becomes fatigued and therefore does not offer long-lasting protection. On the other hand, it only comes into play for low frequencies, below 1 kHz; its contraction occurs too late during sudden noises such as an explosion because the *reflex latency* time (physiological reaction time linked to the human information processing chain) is close to 30 ms.

The inner ear is made up of two sensory organs: the *vestibule*, the balance organ, and the *cochlea*, the hearing organ.

The latter is a hollow snail-shaped bone making 2.5 turns around its axis with three chambers: the *vestibular duct*, the *tympanic duct* and the *cochlear duct* in the center.

The vibrations exerted by the stapes on the oval window are transmitted through the oval window to the liquid called the *perilymph*, rich in sodium, which is contained in the vestibular duct in connection with the tympanic duct at the cochlear *apex* (vertex), in a region called the *helicotrema*. The round window compensates for fluid expansion.

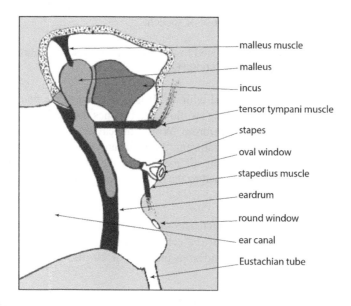

Figure 1.6. *The middle ear in detail*

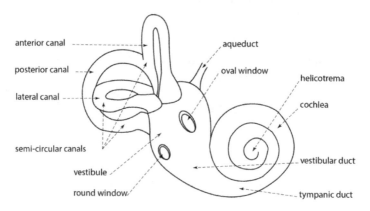

Figure 1.7. *Inner ear*

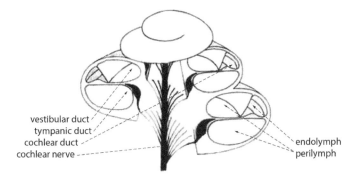

Figure 1.8. *The cochlea*

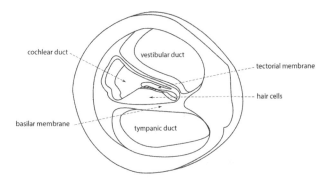

Figure 1.9. *Cross section of the cochlea*

The cochlear duct is closed and filled with *endolymph*, which is rich in potassium. It is separated from the tympanic duct by the *basilar membrane,* on the internal face of which is the sensory organ, containing the receptors, called the *organ of Corti*. These receptors are *ciliated cells*. The inner hair cells are arranged in one row and the outer cells in three rows. The cilia (or *stereocilia*) of the outer cells have their apices fixed to the *tectorial membrane*. The inner and outer cells are connected to the fibers of the *auditory nerve*. The movements of the liquids contained in the ducts cause a deformation of the basilar membrane, which leads to an inclination and a twisting of the cilia connected to the tectorial membrane, which remains fixed. It is this inclination and this torsion that code the sound vibrations through the movement of ions, which polarizes or depolarizes the membrane of each cell.

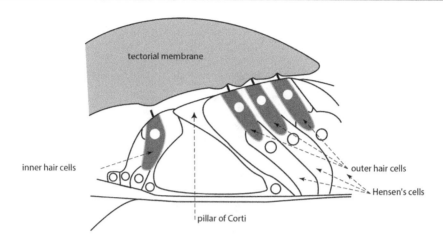

Figure 1.10. *The organ of Corti*

The outer hair cells are selective amplifiers, and the inner cells are true sensory cells. Along the cochlear duct, there is a *tonotopy* (representation of the auditory spectrum), that is, at each of its points a characteristic resonant frequency can be determined. The low-frequency (bass) resonators are located toward the apex and the high frequency (treble) resonators at the base of the duct.

In addition to air conduction of sound vibrations, there is another type of conduction called *bone conduction*. The sound message is directed to the inner ear via the bones of the skull. By this means, it is necessary to bring much more energy to have a stimulation equivalent to aerial conduction. Indeed, we measure that there is at least 30–60 dB of attenuation, depending on the frequency.

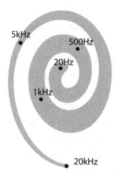

Figure 1.11. *Cochlear duct tonotopy and frequency distribution*

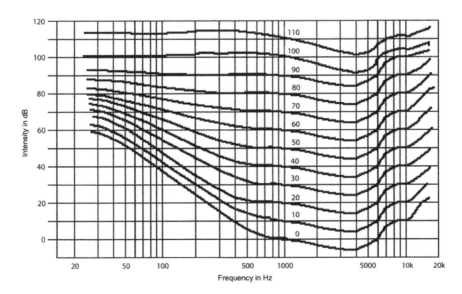

Figure 1.12. *Fletcher and Munson diagram (sound level curves)*

1.2.5.2. *The Fletcher diagram*

Our ear does not have a linear sensitivity with respect to acoustic pressure, that is, depending on the frequency, a signal emitted with the same intensity will seem stronger or weaker to us. Fletcher and Munson's (1933) diagram highlights this phenomenon. Below a certain power and depending on the frequency, it is impossible to perceive a sound. This is what corresponds to our hearing threshold. It is the same for high-powered sound messages that become unbearable, and this is the pain threshold. Fletcher created a diagram that links the frequency plotted on the *x*-axis and the sound power (sound pressure) on the *y*-axis. The curves indicate an equal sound sensation in *binaural listening* (with both ears). This work was standardized in 1961 and defines what are known as *loudness level contours* (Fletcher-Munson *sound curves*).

1.2.5.3. *Spatial listening*

Our ear can locate a sound source quite precisely. Several parameters are responsible for this ability.

It was in 1907 that Lord Rayleigh[6] highlighted principles related to time differences and *interaural level*.

The *Interaural Time Difference* (ITD) is a device characterizing the difference in time that elapses between the arrival of sound waves at each of the ears of a listener. If the source is directly in front, the difference is zero.

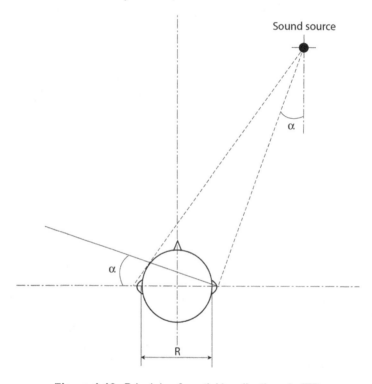

Figure 1.13. *Principle of spatial localization via ITD*

An approximation of the ITD can be obtained by the equation:

$$\Delta_t = \frac{R(\alpha + sin\alpha)}{c}$$

---

6 John William Strutt Rayleigh was a British physicist born in Langford Grove (Essex) (1842–1919). He won the Nobel Prize in Physics in 1904 for his discovery, with William Ramsay, of the inert gas argon and did extensive research on wave phenomena.

where:

- Δ_t is the ITD in seconds;
- R is the radius of the head in meters (8.75 cm by default);
- α is the angle of incidence in radians;
- c is the speed of sound (340 m/s).

For example, for a sound source placed at 30°, we obtain:

$$\Delta_t = \frac{0.0875\,(0.523 + 0.5)}{340} = 2.63.\,10^{-4} s$$

$$= 263\ \mu s.$$

The sound will reach the listener's ears with a time difference of 263 μs. This difference can be likened to a phase shift, which will be analyzed and then interpreted by our brain in order to position the sound source. The maximum ITD is approximately 673 μs.

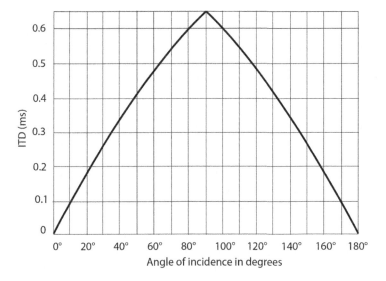

Figure 1.14. *ITD calculation chart*

This phenomenon is also felt for low frequencies, below 1,500 Hz. Above this, it is the *Interaural Level Difference* (ILD), *Interaural Intensity Difference* (IID) or *Interaural Pressure Difference* (IPD) that intervenes.

When sound is emitted from a source closer to one ear than the other, a difference in sound intensity or sound pressure is created. Our auditory system uses this to locate the sound source.

Kendall and Puddie Rodgers (1981) proposed a simple equation for the calculation of the ILD:

$$\Delta_l = 1 + \left(\frac{f}{1{,}000}\right)^{0.8} \times \sin\alpha$$

where:

– Δ_l is the ILD;

– f is the frequency in kHz;

– α is the angle of incidence in radians.

The use of ITD and ILD does not provide all the information to define the position of a sound source. In some cases, the ITD and the ILD may be identical, yet the emission sources are positioned at different locations, as shown in Figure 1.15.

Another parameter, based on the diffraction related to the morphology of the head, is also a localization factor of a sound source. It removes the ambiguity on the localization errors presented previously. It is, today, the most important parameter, which many researchers are working on. The *Head Related Transfer Functions* (HRTF) method concretizes this work.

To understand how HRTF localization works, imagine that when a sound source is placed to the right of an individual, the eardrum of their right ear will receive the sound message, following an approximately rectilinear trajectory. On the other hand, for their left ear, the sound waves will have to take a much more complex path by going around the head to reach the left ear, where, after multiple reflections and diffractions in the auricle and the ear canal, the sound vibrations will hit the eardrum.

As it travels, the timbre of the sound wave, and therefore its spectrum, is modified. These modifications depend on the location of the source within three-dimensional space and are interpreted by the brain in order to determine this location. Note that HRTF localization can determine vertical position, unlike the ITD and ILD.

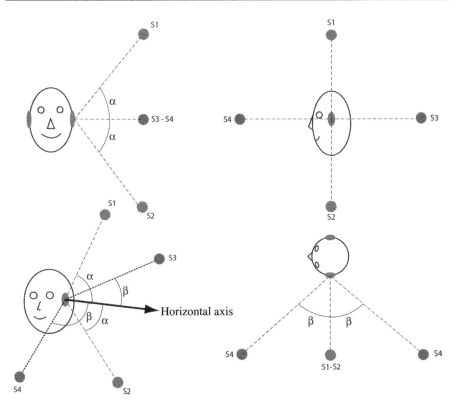

Figure 1.15. *Angle location error in horizontal and vertical field. Sources S1 and S2 (vertical plane) have the same ITD and ILD values as sources S3 and S4 (horizontal plane). For a color version of this figure, see www.iste.co.uk/reveillac/synthesizers1.zip*

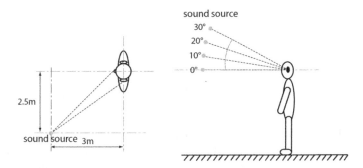

Figure 1.16. *Implementation of an HRTF function. The sound source is placed in front of the observer, and its elevation varies from 0 to 30° in relation to the horizontal plane of their ears*

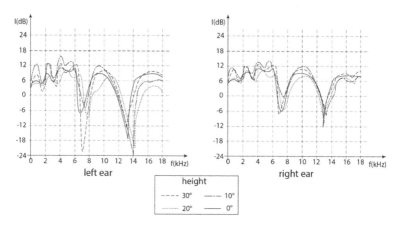

Figure 1.17. *Vertical discrimination by the HRTF method unlike the ITD and ILD methods. Curves differ with height of the sound source in relation to the listener's ear*

It is assumed that the localization angle errors[7], using all the methods (ITD, ILD, and HRTF), are those defined in Figure 1.18.

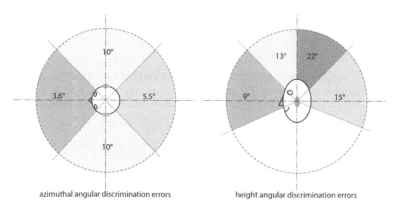

Figure 1.18. *Values of angular discrimination errors*

It should be noted that in the preceding passages, the localization functions studied, ITD, ILD and HRTF, use binaural listening (with both ears); however, localization is possible with monaural listening. It is the auricle of the outer ear that filters the sound because of reflections linked to the angle of incidence of the source. These produce delays and timbre distortions that will be used to determine the origin of the sound.

---

7 According to Blauert's measurements and experiments.

When listening, specific effects can disturb or improve the spatial localization of a sound source. There are two main effects: the *cocktail party effect* and the *precedence* or *Haas effect*.

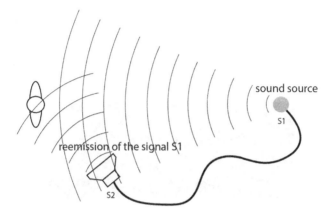

Figure 1.19. *Highlighting the precedence or Haas effect. If the difference between source S1 and S2 is less than 50 ms, the listener does not perceive any delay (no echo)*

The cocktail party effect occurs when we find ourselves in a noisy environment. We can locate a sound source if it is known to us.

The Haas[8] effect highlights the perception of reflected sound compared to direct sound. Our ear does not distinguish between direct sound and reflected (or reemitted) sound when they are separated by less than 50 ms, even if the reflected source is several dB higher than the direct source.

Beyond this duration, the ear no longer perceives a fusion of sound sources but an echo. This phenomenon implies that our auditory system interprets the direction of an acoustic phenomenon as being that of the first source heard. This effect is also called the first wavefront law.

1.3. Sound typology

Sounds, like all the elements present around us, can be classified according to a typology based on characteristics that physics or listening can distinguish.

---

8 According to a study conducted by Helmut Haas in 1951.

1.3.1. *Sounds and periods*

It is rare, except in desired circumstances, that in our natural environment we are in the presence of pure sounds of constant frequencies. The sounds that surround us are often complex, and we can define a sound typology that will allow us to classify them. First, let us look at their periodicity.

A sound is said to be *periodic* if its frequency does not change over time.

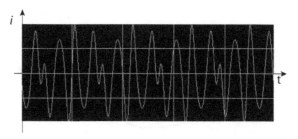

Figure 1.20. *Representation of a periodic sound*

A sound is *aperiodic* when, over time, it is characterized by numerous changes in frequency and amplitude. Most of the noises we hear in our environment are aperiodic.

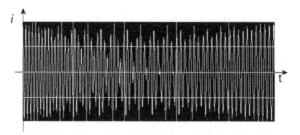

Figure 1.21. *Representation of an aperiodic sound*

White noise is the extreme example of an aperiodic sound because it uses the entire spectrum of audible frequencies.

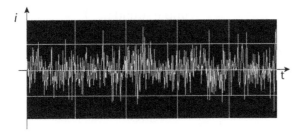

Figure 1.22. *Representation of white noise*

A sound with a very short duration is said to be *impulsive*, unlike a longer sound which is said to be continuous.

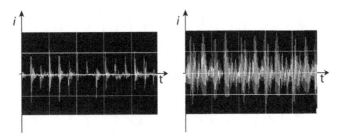

Figure 1.23. *Representation of impulsive and continuous noise*

1.3.2. *Simple and complex sounds*

Simple sounds have sine waveforms, and others are considered *complex sounds*. A complex sound is composed of at least two sine waves. A complex sound is therefore a combination of several simple sounds.

To find the simple waves that make up a complex sound, we use a mathematical tool called the *Fourier[9] transform*.

The Fourier transform breaks down a complex sound into a series of simple sounds (sine waves). A complex real-world sound may contain dozens of simple sine wave components.

---

9 Jean Baptiste Joseph Fourier (1768–1830) was a French mathematician and physicist known for having determined, among other things, the calculation method from a function to a corresponding trigonometric series and vice versa, called the Fourier transform.

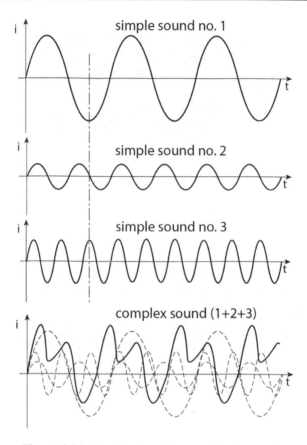

Figure 1.24. *Combination of several single sounds*

The amplitude of a complex sound corresponds, at a given moment, to the sum of the amplitudes of the simple sounds that compose it.

The frequency of a complex sound is equal to the lowest of the frequencies of the simple sounds that compose it.

The frequency of a complex sound is called the *fundamental frequency* (F_0). If the sound is not periodic, it does not implicitly have a fundamental frequency and is instead defined as noise.

1.4. Spectral analysis

Sound spectral analysis brings together several analysis techniques that provide the characteristics of an audio signal. Its results are often presented in the form of graphs.

1.4.1. *The sound spectrum*

The spectrum of a sound is the representation of its analysis in amplitudes and frequencies. The representations used previously always show a sound according to an amplitude variation defined by a vertical *y*-axis along a horizontal *x*-axis, which symbolizes time.

The spectrum of a sound wave does not contain time information.

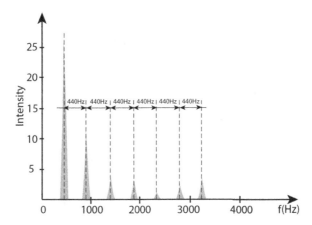

Figure 1.25. *Spectrum of a periodic complex sound*

A spectrum is often represented in the form of a graph with vertical bars (or lines), each corresponding to the amplitude of a given frequency. The lowest frequency represents the fundamental frequency.

This type of graph is particularly suitable for periodic sounds because only certain frequencies are present. These can be a *discontinuous spectrum, line spectrum* or *comb spectrum*.

The other frequencies are called *harmonics*. They are multiples of the fundamental frequency.

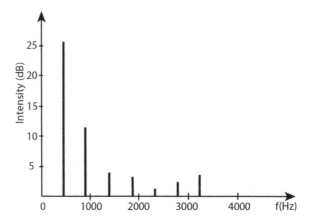

Figure 1.26. *Sound line spectrum of Figure 1.22 (intensity in dB)*

Name	Symbol	Frequency
Fundamental Frequency or First Harmonic	h_1	440 Hz
Second Harmonic	h_2	$2 \times F_1 = 2 \times 440 = 880$ Hz
Third Harmonic	h_3	$3 \times F_1 = 3 \times 440 = 1\,320$ Hz
Fourth Harmonic	h_4	$4 \times F_1 = 4 \times 440 = 1\,760$ Hz
Fifth Harmonic	h_5	$5 \times F_1 = 5 \times 440 = 2\,200$ Hz
Sixth Harmonic	h_6	$6 \times F_1 = 6 \times 440 = 2\,640$ Hz
Seventh Harmonic	h_7	$7 \times F_1 = 7 \times 440 = 3\,080$ Hz

Table 1.2. *The harmonics of (A) 440 Hz*

Harmonics 2, 4, 6 and so on are called even harmonics, and 3, 5, 7 and so on are called odd harmonics.

In nature, pure sounds are almost non-existent; they are always accompanied by information that allows for their identification. It is here that, together with the fundamental sound, harmonics and inharmonic overtones (also called partial tones) appear.

The fundamental frequency of the sound defines the general pitch of the sound signal; the harmonics, which are multiples of the fundamental frequency of the sound, can be even or odd. Even harmonics are more pleasing to the ear. Inharmonic overtones form the remainder of the signal that are not harmonics.

The combination of the fundamental sound, the harmonics and the non-harmonic sounds form the sound spectrum of a signal. It is therefore necessary to fully understand their role to create an artificial sound, especially if it must resemble a real sound.

Instead of a bar graph, one often needs to visualize the distribution of frequencies in the spectrum by using a graph that shows the *spectral envelope* of the sound.

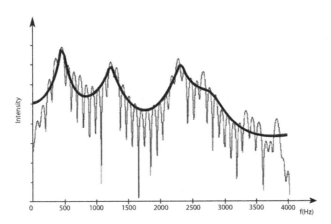

Figure 1.27. *Spectral envelope of an aperiodic complex sound*

This type of graph is excellent for aperiodic sounds with many frequencies. Sounds like this are said to have a *continuous spectrum*.

1.4.2. Sonogram and spectrogram

One of the disadvantages of the spectral representation, as we saw earlier, is that it does not include any temporal information. However, when the signal is *pseudo-periodic* (a sequence of periodic signals over time: musical phrase, speech, etc.), it can be useful to take time into account.

To meet this demand, scientists have developed types of graphs with three parameters (frequency, intensity and time), called *sonograms* or *spectrograms*.

The sonogram is a 2D graph in which the sound intensity is defined by a scale of colors or shades of gray. The spectrogram is usually a 3D graph in which time and frequency are placed on the x- or z-axes and the loudness on y.

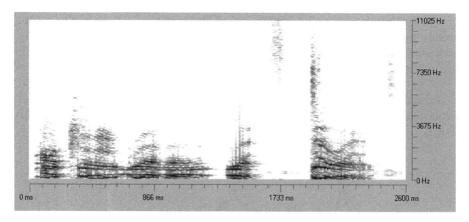

Figure 1.28. *Sonogram of a sound sequence*

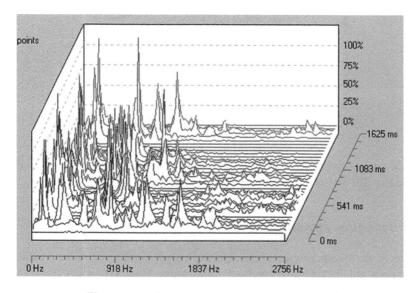

Figure 1.29. *Spectrogram of a sound sequence*

1.5. Waveforms

When considering a waveform or the shape of an audio signal, the frequency is not the most important parameter; regardless of the pitch, 1 Hz, 10 Hz, 5,000 Hz or 100 kHz, the shape of the signal remains the same. In the context of this book, we will study the waveforms in the audible domain.

The most common shapes are sine waves, square waves, rectangular waves, triangular waves and sawtooth waves.

1.5.1. *Sine wave*

The sine (or sinusoidal) wave is the origin of all waveforms; it represents pure sound, with no harmonics or inharmonic overtones (see section 1.4.1). Its frequency determines its pitch.

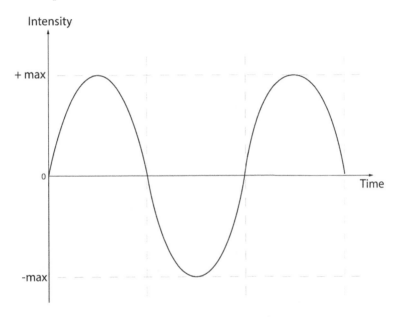

Figure 1.30. *Representation of a sine wave*

1.5.2. *Square wave*

Although the sine wave is considered the most important, strangely it is the square wave whose shape is the simplest to produce electronically. A square wave, as its shape shows, only exists in two states: low and high.

A square wave is symmetrical when it spends as much time above its equilibrium point (0 in Figure 1.30) as below, otherwise it is asymmetrical.

We could say, by analogy, that the action of opening or closing a switch produces an asymmetrical square wave; when high, current flows; when low, there is no current.

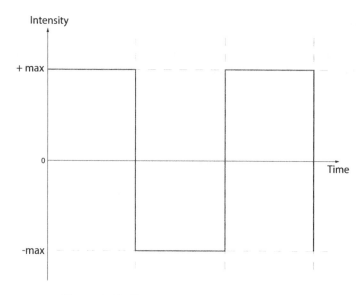

Figure 1.31. *Representation of a square wave*

It contains the fundamental sound with odd harmonics. The sum of these added to the fundamental frequency makes it possible to obtain the square shape. When heard, a square wave generates a soft and hollow sound, similar to a woodwind instrument like the clarinet.

Harmonic intensity I, of this type of wave, decreases according to the formula

$$I = \frac{1}{f}$$

where f is the value of the frequency concerned.

By deduction, we can say that the intensity decreases by $1/n$ where n represents the odd harmonic number.

A perfect square wave, like the one in Figure 1.31, would contain infinite odd harmonics.

By default, a square wave is symmetrical, half of its intensity being distributed on either side of the zero point.

By applying the Fourier transform (see section 1.3.2), we can see that a square signal is made up of the addition of several simple sinusoidal signals.

Figure 1.33 shows the addition of six sinusoidal signals of the same phase, a fundamental frequency, $h1$, and five odd harmonics $h3$, $h5$, $h7$, $h9$ and $h11$ decreasing by an intensity of $1/f$.

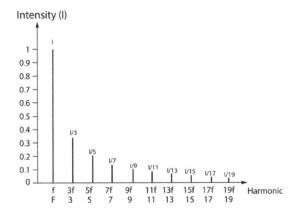

Figure 1.32. *Harmonic spectrum of a square wave*

1.5.3. Rectangular wave

A rectangular wave is a variation of the square wave. Similarly, it has only two states, up and down. The difference lies in the division of its period whose duration in the high state differs from that of the low state, as shown in Figures 1.34(a) and (b).

In fact, a rectangular signal is characterized by T, its period, and T_H which corresponds to the duration during which it is at its highest level, that is to say, at its maximum intensity.

This makes it possible to define what is called the duty cycle r, which is equal to:

$$r = \frac{T_h}{T}$$

r is a unitless value that is between 0 and 1 (or between 0 and 100%).

32   Synthesizers and Subtractive Synthesis 1

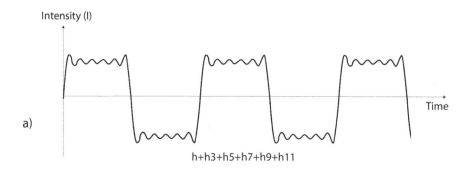

a)

h+h3+h5+h7+h9+h11

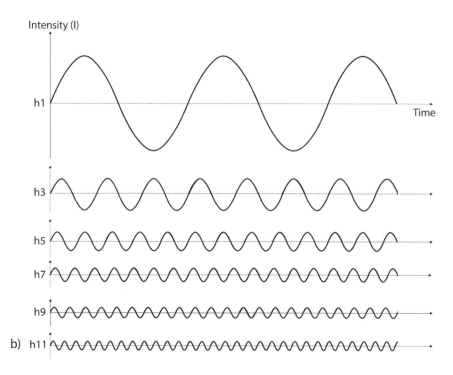

b)

Figure 1.33. *(a) Construction of a square signal and (b) from a fundamental frequency and five odd harmonics*

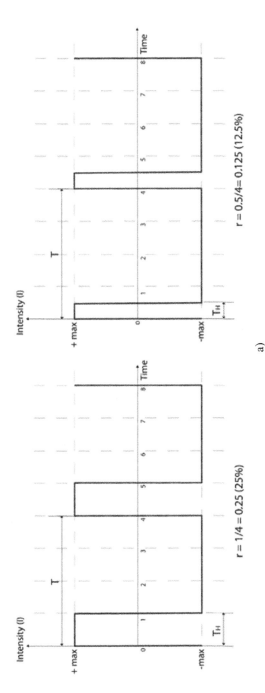

Figure 1.34a. Four possible representations of a rectangular wave

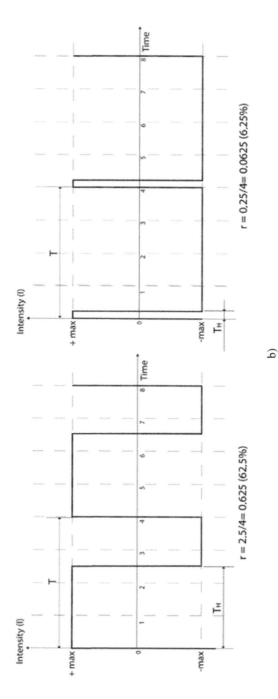

Figure 1.34b. *Four possible representations of a rectangular wave (continued)*

REMARK.– If the duty cycle $r = 0.5$, the rectangular wave is identical to a square wave.

A rectangular signal produces a rather nasal sound, although this aspect depends more or less on the duty cycle.

1.5.4. *Pulse wave*

A pulse wave (or pulse train) is an asymmetrical wave, also derived from the square wave. It also alternates between two states, high and low. In this case, the low state equals zero intensity. Like the rectangular wave, it has a duty cycle.

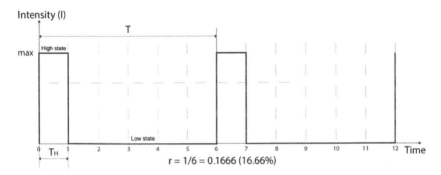

Figure 1.35. *Representation of a pulse wave*

The use of a pulse signal is very often encountered in subtractive sound synthesis through Pulse Wave Modulation (PWM) to create, for example, a chorus effect.

1.5.5. *Triangular wave*

This wave is like a square wave and also contains odd harmonics. However, they decrease according to the formula $1/f^2$ with f representing the chosen frequency. In this type of wave, the fundamental frequency is present, but the attenuation of the harmonics is very fast.

36 Synthesizers and Subtractive Synthesis 1

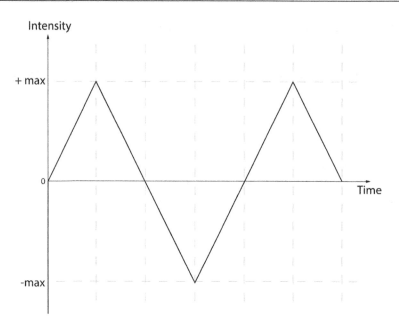

Figure 1.36. *Representation of a triangular wave*

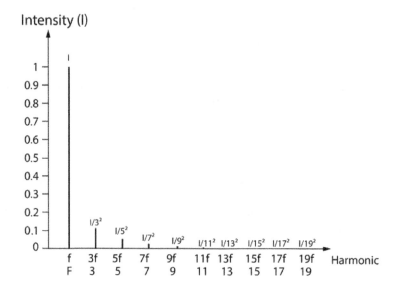

Figure 1.37. *Harmonic spectrum of a triangular wave*

A perfect triangular signal, like the one in Figure 1.36, would contain infinite odd harmonics.

By deduction, we can say that the intensity I decreases by $1/n^2$ where n represents the odd harmonic number.

Figure 1.38 shows the addition of six sinusoidal signals of the same phase, a fundamental frequency, $h1$, and five odd harmonics $h3$, $h5$, $h7$, $h9$ and $h11$ decreasing by an intensity of $1/f^2$.

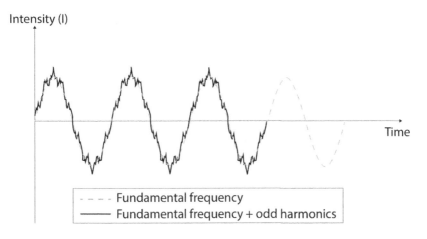

Figure 1.38. *Construction of a triangular signal from a fundamental signal and five odd harmonics*

1.5.6. *Sawtooth wave*

The shape of this wave is obtained by adding all the harmonics to the fundamental frequency, whether they are even or odd. It can have two forms, descending or ascending.

It is a rich and full signal, which sounds very brassy. Its amplitude decreases slowly.

In subtractive synthesis, it is often considered one of the best waveforms for building other sounds, especially those of string instruments or certain brass instruments.

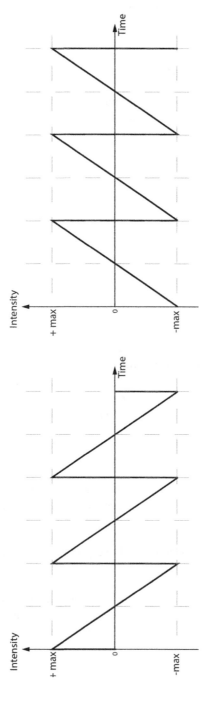

Figure 1.39. (a) Representation of a downward or (b) upward sawtooth wave

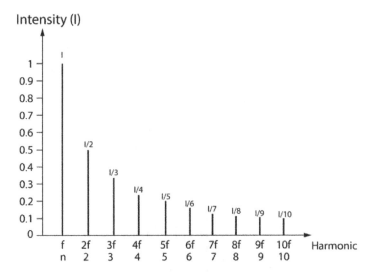

Figure 1.40. *Harmonic spectrum of a sawtooth wave*

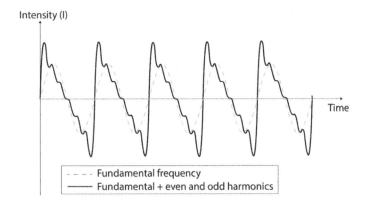

Figure 1.41. *Construction of a sawtooth signal from a fundamental frequency and the first five odd and even harmonics*

1.6. Timbre

The notion of timbre is too difficult to describe in terms of the fundamental frequency, the harmonics or the loudness (sound intensity). Timbre is a complex psychoacoustic concept. Sounds can be strident, dull, dry, hot, etc., the list of adjectives is far from exhaustive.

1.6.1. Transient phenomena

Sound perception exists within a wider context. Each sound begins, stabilizes and then fades. Each single moment in the act of hearing a sound is a *transient phenomenon*. There are several types: *attack transients* and *release transients*, which describe the beginning and the end of sound phenomena.

We can add other parameters to this list, which, like *vibrato* and *tremolo*, can either occur in a single phase (often when the sound has stabilized), or in every phase.

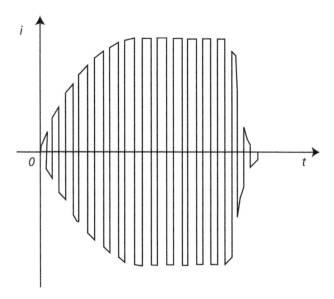

Figure 1.42. *Transients in the sound of a pipe organ*

The attack transient is what gives a sound its unique signature and what allows us to distinguish between the sound of a clarinet and a flute, for example. Its duration varies, typically oscillating between 1 and 100 ms. The human ear needs 40–50 ms to differentiate and recognize a sound. This transient is often complex. In addition to its duration, it is necessary to consider its *slope*, its instability, its spectral composition, the number of components and their order of arrival.

The sustain, or when a sound is produced, is often steady, although the environment and the playing of the musician can modify it.

The extinction transient (or *decay*) is often linked to environmental parameters (*echo, reverberation, damping*, etc.) and playing parameters in the case of an instrumentalist. It can be very short or very long in duration (sound synthesis peripherals or natural or artificial echo phenomena), between 1 and 5,000 ms.

Together the transient phenomena, attack, sustain and decay are often called the *sound envelope*.

1.6.2. Range

Each sound source, especially musical instruments and voices, can emit sound over a certain range of frequencies. This range is also known as the tessitura. However, we must be careful to distinguish between two concepts: the *fundamental range* and the *spectral range*.

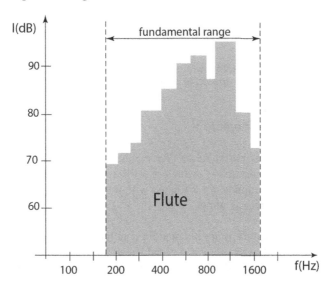

Figure 1.43. *Fundamental range of a flute*

The fundamental range (see Figure 1.43) only includes fundamental frequencies and harmonics. In an ideal setting, this is equivalent to the spectral range, where the range of musical instruments can be simplistically described as a set of fundamental frequencies plus harmonics, with higher ranking harmonics having decreasing intensity. However, it is not always so in reality.

The spectral range (see Figure 1.44) contains the entire sound spectrum that can be emitted by the chosen source.

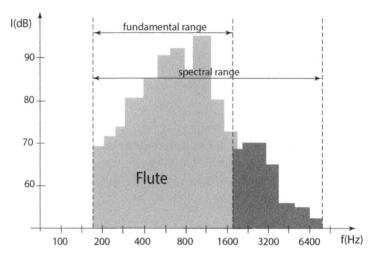

Figure 1.44. *Fundamental range of a flute compared to its spectral range*

1.6.3. *Mass of musical objects*[10]

This notion of mass is attributed to Pierre Schaeffer[11] who introduced it in his book *Traité des objets musicaux*, which is still referenced today. Speaking of concrete sounds, that is, sounds not affiliated with a musical instrument that we are able to culturally recognize, he wrote:

> But if we listen carefully to this random concrete sound (for example, it may come from a membrane, a metal sheet, a rod...), we notice that, without having a clearly locatable pitch like traditional sounds, it

---

10 We take up here the notion dear to Pierre Schaeffer: "The sound object, given in perception, designates the signal to be studied... listen to the sound as a sound event".

11 Pierre Schaeffer (1910–1995) was the founder of the ORTF research department (Office of French Broadcasting and Television), researcher in the fields of audio-visual communication and music, founder of concrete music and GRM (Musical Research Group) in 1951, composer (*Variation sur une flûte mexicaine, Suite pour 14 instruments, Symphonie pour un homme seul, Toute la Lyre, Orphée 51, Masquerage*, etc.) and author of the monumental *Traité des objets musicaux*.

nevertheless presents a sound "mass" situated somewhere in the tessitura and is more or less characterized by occupying fairly decipherable intervals. It includes, for example, several sounds of gradually evolving pitch, crowned or surrounded with a conglomeration of partials that also evolve, all of this more or less locatable in a certain pitch zone. The ear soon manages to locate the most prominent components and aspects of these, provided it is trained to do so; such sounds can then become as familiar to it as traditional harmonic sounds: they have a characteristic mass.

His comments reveal the extent of the complexity involved in attempts to define the timbre. The concept of timbre extends far beyond the physics of the phenomenon, into the realm of psychoacoustics. Schaeffer continues:

It is high time to point out that, when the musician keeps on saying "a very rich note," "a good, or bad, timbre," and so on, it is because he does not confuse two concepts of timbre: one relative to the instrument, an indication of source in ordinary listening [...] and the other relating to every object produced by the instrument, an appreciation of the musical effects contained in the objects themselves, effects that are desired by both musical listening and musical activity. We even went further, attaching the word timbre to a part of the object: the timbre of attack, as opposed to its steepness.

But, defined in this way, the timbre of an object is nothing other than its sound form and matter, a complete description of it, within the limits of the sounds that a given instrument can produce, when all the variations in facture it may have are taken into account. The word timbre with reference to the object is therefore of no further help to us in the description of the object in itself, since it merely involves us in a reanalysis of the most subtle of the informed perceptions we have of it.

1.6.4. *Classification of sounds*

Based on the observations relating to the concept of timbre presented above, we can define a classification of different types of sound message.

Category	Composition	Spectrum	Example
Pure sound	Tonal sound without harmonics	Filiform	Sinusoidal sound (BC generator, synth VCO, etc.)
Tonal sound	Sound with identifiable pitch	Single band	Note played by an instrument (piano, harpsichord, violin, etc.)
Tonal group	Group of multiple tonal sounds	Multiple bands	Chord played by an instrument (piano, harpsichord, organ chord, etc.)
Nodal sound	Aggregate sound without an identifiable pitch	Multiple bands	Set of percussions (several different cymbals together)
Nodal group	Set of multiple nodal groups	Single band	Percussive sound (cymbal)
White noise	Group containing every pitch	Whole spectrum	White noise generated by a synthesizer
Complex or channeled sound	Mixed group containing tonal sounds, tonal groups, nodal sounds, nodal groups	Complex shape	Natural sound (bell, gong, metal sheet, etc.)

Table 1.3. *Classification of sound messages*

1.7. Sound propagation

A sound wave propagates through its surrounding media by means of specific phenomena that result in particular behaviors. We will study the main principles.

1.7.1. *Dispersion*

A sound wave emitted by a point source disperses as a set of concentric spheres.

Sound propagates through gaseous media like air, which is the most common transporting medium, in the form of alternating compressed and dilated layers. The set of wavefronts vibrates in phase (with each other).

In order to describe certain phenomena, scientists introduced the abstract notion of plane waves, which do not actually exist. Plane waves are sections of spherical waves. When the source is far enough away from the point of audition, spherical waves have a large radius of curvature, and so may be thought of as plane waves.

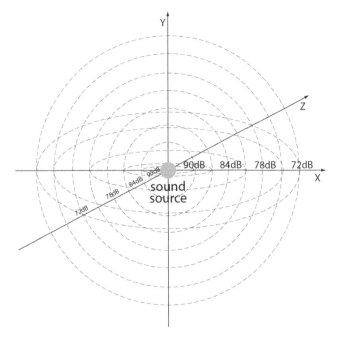

Figure 1.45. *Spherical dispersion of a point sound wave. The sound pressure level is reduced by 6 dB when the distance doubles*

1.7.2. Interference

Figure 1.46. *Interference between two waves on the surface of a liquid*

When two sound waves meet, they overlap, and their intersection creates *constructive* or *destructive interference*.

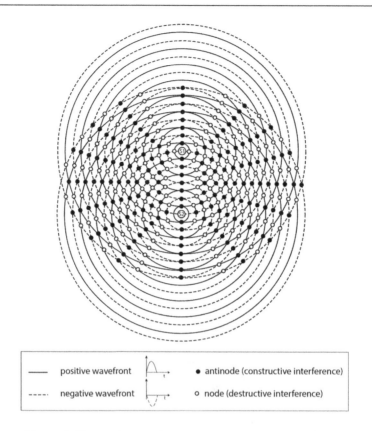

Figure 1.47. *Interference between two identical waves (same frequency and same amplitude). Sources S1 and S2 emit sound waves which are superimposed by creating interferences (nodes and antinodes)*

If you strike a tuning fork and then rotate it near your ear, you will notice that it sounds louder or softer depending on the angle of rotation. This simple experiment demonstrates the creation of constructive and destructive interferences.

When the sound waves emitted by each branch of the tuning fork are in phase, they add together to create constructive interference. But if they are perfectly out of phase, they generate destructive interference and the intensity of the signal is doubled. But if they are perfectly out of phase, they generate destructive interference and the intensity is zero, so no sound is produced.

NOTE.– Between constructive interference (perfect sum of both signals) and destructive interference (perfect cancellation of both signals), there is an intermediate zone in which the amplitude of the signal varies between its maximum and its minimum. This notion is discussed again in section 1.7.9.

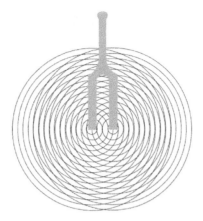

Figure 1.48. *Interference produced by a tuning fork*

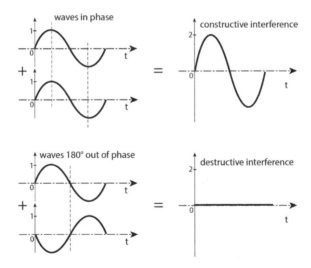

Figure 1.49. *Constructive and destructive interference*

If the sound waves are identical, it is easy to determine the shape of the interference. If not, they can be determined by the *superposition principle*: when two sound waves occupy the same space at the same time, the resulting disturbance is the sum of the different disturbances at each point in space and time.

1.7.3. Diffraction

When a sound wave encounters an obstacle, it goes around it. The edge of the obstacle becomes the center of a new wave called a *secondary wave* or a *diffracted wave*.

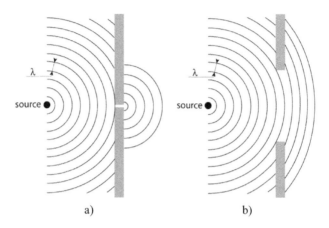

Figure 1.50. *Diffraction of a wave through an opening. (a) The opening is small compared to the wavelength λ, a point source is created; (b) the opening is larger than λ, there is practically no diffraction*

Diffraction is what allows us to hear sounds emitted by a source located behind an obstacle.

Waves do not propagate in straight lines. Once a wave has moved past the obstacle, it again propagates in every direction. This phenomenon was demonstrated by the physicist Christian Huygens[12] who demonstrated this and coined the *Huygens' principle*:

> Each point on a wavefront may be viewed as the point source of a wave propagating in the same direction as the original wave. The next wave front may be obtained by summing all these new waves.

---

12 Christian Huygens, born in La Haye, Holland (1629–1695), was a mathematician, physicist and astronomer, and author of *De ratiociniis in ludo aleae* in 1657 and *Horologium oscillatorium* in 1673. He discovered Saturn's rings, the rotation of Mars, and the Orion Nebula. He also worked and published numerous publications on the wave theory of light, which enabled him, among other things, to explain reflection and refraction.

In simpler terms, a wave can be thought of as a sum of elementary waves vibrating in phase, which are moving in the same direction as the original wave.

When the obstacle placed in front of the sound-emitting source is smaller than the wavelength, it does not exist from the perspective of the wave – the sound wave propagates around it as if it were "invisible". The greater the wavelength-to-width ratio of the obstacle (λ/l), the greater the effect of diffraction. When the ratio is large, an observer located behind the obstacle does not perceive any difference compared to when the obstacle is removed.

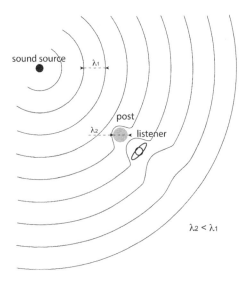

Figure 1.51. *Circumvention of an obstacle smaller than the wavelength*

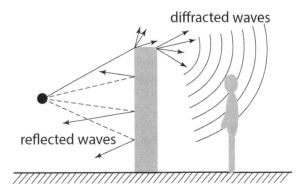

Figure 1.52. *Emission of secondary waves on the edges of an obstacle*

When the width of the obstacle is large, part of the sound wave is reflected, and the edges of the obstacle become secondary propagation sources.

1.7.4. Reflection

When a sound wave encounters an obstacle that it cannot circumvent, because the width of the obstacle is greater than the wavelength of the wave, the wave is reflected. The *reflection* of sound waves follows the appropriately named *law of reflection* (Descartes' law)[13]. This law states that the new direction of the sound wave after hitting a point on the surface is equal to the angle of the wave to the normal at this point at the moment of impact; in other words, the *angle of incidence* is equal to the *angle of reflection*.

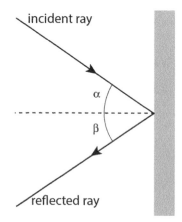

Figure 1.53. *Angles of incidence and reflection of a sound wave ($\alpha = \beta$)*

Note that the reflected waves can interfere with the incident waves to create constructive or destructive interference. Within a thin layer very close to the surface hit by the wave, the sound intensity is increased by the sum of the incident wave and the reflected wave. Here, the sound pressure is doubled, which increases the intensity by about 6 dB.

If the obstacle reflecting the wave has a concave shape, a *focusing* phenomenon occurs. In the opposite case, when the obstacle has a convex shape, a *scattering* phenomenon occurs.

---

13 René Descartes, born in Touraine, France (1596–1650), was a philosopher, physicist, mathematician and author of the famous *Discours de la méthode*, *Traité du Monde et de la Lumière*, *La Dioptrique* and *Météores*.

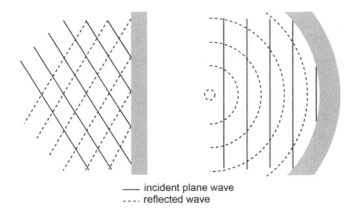

Figure 1.54. *Reflection of a sound wave off a flat surface and a concave surface*

The wavefronts that hit a wall create reflected waves as if there were an "image" of the sound source located at the same distance on the other side of the wall.

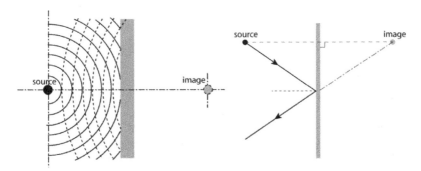

Figure 1.55. *Reflection of a sound wave off a wall and image of its source*

When a sound wave strikes a hard surface, there is no phase change in the reflected wave. We say that the surface has an acoustic impedance greater than air. But when a sound wave encounters an obstacle with a lower impedance, there is a phase inversion (e.g. when a sound transmitted by a solid meets the air).

1.7.5. Reverberation (reverb)

A sound wave, when it propagates in the air, is reflected off obstacles, walls, floors, ceilings and various objects that it encounters. These reflections follow one another and cause the sound wave to travel longer and longer paths while decreasing in energy as it is absorbed by the materials it strikes.

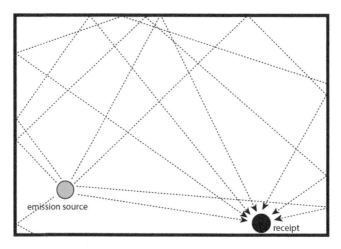

Figure 1.56. *Possible trajectories (in dotted lines) of the reflections of a sound wave causing reverberation*

The number of reflections is usually large since the speed of sound (around 340 m/s) is typically large relative to the size of the location where the sound is propagating. For a room 10 × 10 m in size, we can estimate the number of reflections to be around 30 per second.

Listeners in a reverberating environment will hear the same sound signal repeatedly at slightly different times. This stretches the sound, making it last for longer than the sound that was originally emitted. This phenomenon is known as *reverberation (reverb)*, and the additional duration by which the sound signal is extended is called the *reverb time*.

Reverb often makes sounds seem more lingering (in large spaces like a church, cathedral, etc.) and/or louder (in small venues without much sound absorption) than sounds that are played outdoors.

In small rooms, the time differences are small enough that the sound signal does not repeat itself, and the listener perceives it as a single sound. If the dimensions of the space in which the sound is emitted are large enough (like a valley) or are concave shaped (like a tunnel), the sound signal takes much long to return to the listener, who can then distinguish between each separate instance (repetitions). This is known as an *echo*, which is a special case of reverb.

In order to measure reverb, the field of acoustics defines the theoretical reverb time of a given location as the time taken for the sound energy to drop to one-millionth of its initial value, which corresponds to an attenuation of 60 dB.

Architects often analyze the reverb time when designing buildings. Sabine's equation[14] can be used to calculate the theoretical reverb time:

$$T = \frac{0.163 \times V}{A}$$

where:

– 0.163 is the constant in s/m;

– T is the reverb time in seconds;

– V is the volume of the room in m³;

– A is the absorption surface area in m² (sum of the product of the area of each surface S_n in m² and its Sabine absorption coefficient α_n):

$$A = (\alpha_1 \times S_1) + (\alpha_2 \times S_2) + \cdots .$$

Table 1.4 presents some Sabine absorption coefficients, according to the frequency of the sound waves emitted.

As α_n tends toward 1, the wall absorbs an increasing proportion of the sound energy. If no energy is reflected, the material is said to be perfectly absorbent. As α_n tends toward 0, the material becomes increasingly reverberant.

Sabine also showed that (Sabine's laws):

– the curves describing the build-up and decay of a sound are essentially exponential and are complementary, meaning that the increase in sound energy over a given period is equal to the decrease over the same period;

– the effect of an absorbent material does not depend on its position;

– the release period of a sound is approximately the same at every point in the sound space;

– the reverb is independent of the position of the sound source within the space.

---

14 Wallace Clement Sabine (1868–1919), born in Colombus, OH, was an acoustics engineer who graduated from Harvard.

However, these laws need to be nuanced in some cases. For example, reverb works in a slightly different way in the presence of obstacles, people and convex or concave walls.

Material	125 Hz	250 Hz	500 Hz	1 kHz	2 kHz	5 kHz
Wood	0.09	0.11	0.1	0.08	0.08	0.1
Plaster cement	0.01	0.01	0.02	0.02	0.02	0.03
Wallpaper	0.01	0.02	0.04	0.1	0.2	0.3
Carpet	0.12	0.2	0.25	0.45	0.4	0.35
Fiberglass: thickness 4 cm	0.3	0.7	0.88	0.85	0.65	0.6
Raw plaster	0.04	0.03	0.03	0.04	0.05	0.08
Painted plaster	0.01	0.01	0.02	0.03	0.04	0.05
Raw cement	0.01	0.01	0.01	0.02	0,05	0.07
Plywood 5 mm + 5 cm air	0.47	0.34	0.3	0.11	0.08	0.08
Floor tile	0.01	0.015	0.02	0.025	0.03	0.04
Ordinary glazing	0.3	0.22	0.17	0.14	0.1	0.02

Table 1.4. *Sabine absorption coefficients*

1.7.6. *Absorption*

When a sound wave hits an obstacle, it is reflected and loses some of its energy. This energy is absorbed by the material. The ability of a material to capture or absorb sound energy is described by its *absorption coefficient*, often denoted α:

$$\alpha = 1 - |r|^2$$

where:

– α is the absorption coefficient (note when $\alpha = 1$: perfectly absorbent; $\alpha = 0$: perfectly reflective);

– r is the reflection factor.

Hard materials such as marble, cement and plaster have absorption coefficients ranging from 0.01 to 0.05, whereas porous or fibrous materials such as carpet, felt, or fiber glass have coefficients between 0.2 and 0.4 (see Table 1.4).

1.7.7. Refraction

When a sound wave passes from one environment to a different environment, it changes speed (*discontinuous celerity*) and direction. This generates both a *reflected wave* and a *refracted wave* (especially in the case of thin materials) with a lower energy than the original incident wave.

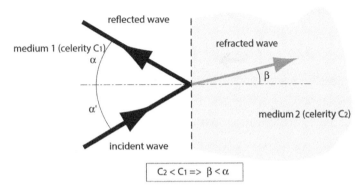

Figure 1.57. *Refraction of a sound wave (α = α')*

If the propagation speed in the first medium is greater than the speed in the second medium, the angle of refraction is smaller than the angle of incidence or the angle of reflection.

1.7.8. Doppler effect

The physician Johann Christian Doppler[15] discovered the Doppler Effect in 1845, the effect arising from the receiver's perception of the offset between the emitted wave and the received wave whenever the emitter and the receiver are moving relative to each other.

I am sure you have noted the sound made by a passing car when you are standing at the edge of the road. The sound is higher pitched when the car is moving closer, and lower pitched when it is moving away.

Whenever a sound source moves toward an observer, the sound waves are *compressed*, reducing the wavelength λ_2 and increasing the frequency of the sound. The perceived sound is therefore higher. Whenever the source moves away from the

---

15 Johann Christian Doppler (1803–1853) was an Austrian physicist.

observer, the opposite phenomenon occurs. The sound waves *dilate*, the wavelength λ_1 increases and the frequency decreases, making the sound appear lower.

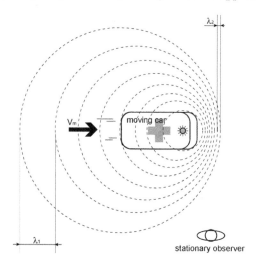

Figure 1.58. *Representation of the Doppler effect*

$$\lambda_1 = (c + V_m)T$$

$$\lambda_2 = (c - V_m)T$$

where:

- c is the speed of sound (m/s);
- λ_1 is wavelength 1 (m);
- λ_2 is wavelength 2 (m);
- V_m is the speed of the moving object (m/s);
- T is the period (s).

One simple way to visualize the Doppler effect is to imagine yourself standing in the water at the seaside. Suppose that the waves hit your feet every 5 s. If you run further into the sea, the waves will hit you more often as you run, and so their frequency is higher (compression). If you turn around and run back to shore, the waves will hit you less often as you run, and so their frequency is lower (dilation). Note however that moving around in the water does not change the total number of waves.

1.7.9. *Phase and beat*

Phase is a phenomenon that is measured in degrees. However, for a musician, it is considered a time gap between two waves.

In Figure 1.59, we can say that:

– the phase shift between waves a and c is 90° with a time gap of t2–t0;
– the phase shift between waves b and c is 45° with a time gap of t2–t1.

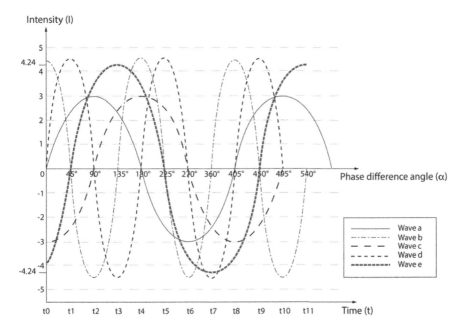

Figure 1.59. *Waves and phase difference. For a color version of this figure, see www.iste.co.uk/reveillac/synthesizers1.zip*

If we add wave *a* with an intensity of I_1 and wave *c* with an intensity of I_2 of the same frequency, we obtain wave *e*. Its intensity I_3 is equal to:

$$I_3 = \sqrt{(I_1 \cdot \sin\alpha_1 + I_2 \sin\alpha_2)^2 + (I_1 \cdot \cos\alpha_1 + I_2 \cdot \cos\alpha_2)^2},$$

That is:

$$I_3 = \sqrt{(3 \cdot \sin(0) + 3 \cdot \sin(90))^2 + (3 \cdot \sin(0) + 3 \cdot \sin(90))^2},$$

$$I_3 = \sqrt{(3.0 + 3.1)^2 + (3.1 + 3.0)^2},$$

$$I_3 = \sqrt{3^2 + 3^2},$$

$$I_3 = \sqrt{18} = 4.24.$$

The phenomenon known as a *beat* occurs when two sounds have similar frequencies. It is a direct consequence of the sensitivity of our auditory perception system. The alternating constructive and destructive interference of the sounds makes the volume of the sound signal seem to alternate between loud and quiet.

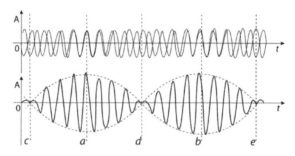

Figure 1.60. *Beat. In a and b, the signals are in phase; in c, d and e, they are totally out of phase*

The frequency of the beat is equal to the absolute value of the difference between the two frequencies of the sound waves:

$$f_{beat} = |f_2 - f_1|.$$

If the frequency of the beat is greater than 50 Hz, our brain can distinguish between the two sound sources. If not (<50 Hz), our brain detects a sound with an intermediate pitch (often called the *subjective tone* or *combination tone*) with an intensity that appears to vary at the rate of the beat frequency.

1.8. Noise

As mentioned in section 1.3.1, noise is often used in sound synthesis. It is a very interesting aperiodic or non-periodic sound; it has no harmonic structure and is associated with colors to distinguish each of its timbres or shapes.

1.8.1. *White noise*

White noise is composed of all audible sound frequencies at equal power. It is an analogy to white light, which mixes all frequencies of the light spectrum.

From one octave to another, the sound frequencies double, that is, the intensity increases by 3 dB per octave.

White noise is similar to the sound of air conditioning blowing in a bedroom, the sound of the radio on an unused frequency, the sound of a loud waterfall or the sound of a fan running in the background of a room.

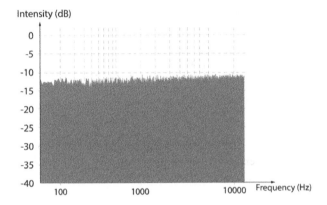

Figure 1.61. *White noise spectrum*

It is very monotonous and remains at an even intensity. It is often used to mask unwanted noise from the environment.

This type of noise is used to treat tinnitus or even to help babies to sleep.

1.8.2. *Pink noise*

Pink noise is flat when defined in logarithmic space. This is one of its main characteristics, and its intensity is the same over intervals ranging from 30 to 60 Hz and from 3,000 to 6,000 Hz. Indeed, these intervals are proportional since they correspond to a 50% increase in frequency.

In music, the frequency band covered by the first octave, from 32.7032 Hz to 61.7354 Hz, or 29.0322 Hz, is much narrower than the last octave that goes from 2,093 to 3,951.07 Hz, that is 1,858.07 Hz, which leads to a much greater sound intensity in the higher octaves. Pink noise takes this phenomenon into account, which is why it is often used as a reference signal to calibrate audio systems in engineering.

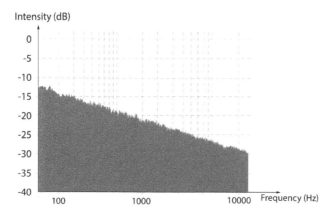

Figure 1.62. *Pink noise spectrum*

The sound power of pink noise, compared to that of white noise, decreases by 3 dB per octave. Its power density is proportional to $1/f$ where f is the frequency. Therefore, pink noise is also called $1/f$ noise.

1.8.3. *Red noise*

Red noise is also called *Brown noise, Brownian noise, random walk noise* or even *drunkard's walk noise*. Its intensity decreases by 6 dB per octave, which implies that its density is proportional to $1/f^2$ where f is the frequency. Its power decreases as its frequency increases.

It is described as red because of the shape of its sound spectrum, which is between pink noise and white noise.

As for the term "brown", it has nothing to do with the light spectrum or a color, but with the expression of *Brownian motion*, which is defined as the continuous agitation of pollen molecules in water, a phenomenon discovered in 1827 by the Scottish botanist Robert Brown[16], whose simulation algorithm produces this noise.

It is a very rich signal in low frequencies, as can be seen in Figure 1.63.

---

16 Robert Brown (1773–1858) was a Scottish surgeon, botanist and explorer.

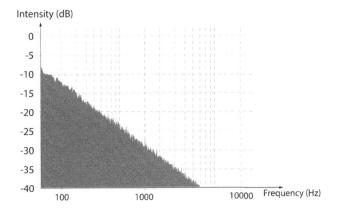

Figure 1.63. *Red noise spectrum*

1.8.4. *Blue noise*

Even if *blue noise* is encountered less frequently than the previous noises in sound synthesis, some synthesizers offer it, such as the MatrixBrute from Arturia, for example. It is also called azure noise.

Blue noise increases in intensity with increasing frequency. Its intensity density increases by 3 dB per octave.

One of the applications of blue noise in sound is *dithering*, a process used by sound engineers to smooth out sound while reducing the audibility of distortion.

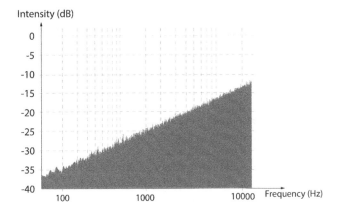

Figure 1.64. *Blue noise spectrum*

It is a noise that is quite hard on the ear, its energy mainly being concentrated in the high frequencies. It is reminiscent of the hissing sound of water flowing from a bent garden hose.

1.8.5. *Purple noise*

Also called violet noise, it is a cousin of blue noise, more intense and violent. Its energy is even more concentrated in the high frequencies. It increases by 6 dB per octave when the frequency increases.

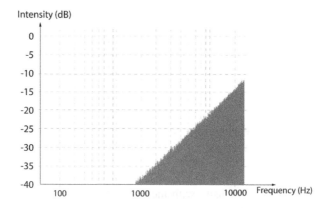

Figure 1.65. *Purple noise spectrum*

1.8.6. *Gray noise*

It is very similar to pink noise because it sounds the same at all frequencies. Looking at its curve, as shown in Figure 1.66, we can see that it represents the psychoacoustic curve of equal A-weighted loudness, which allows us to obtain equal power for each frequency (see section 1.2.2 on the dB(A) and the psophometric curve).

Gray noise was designed to match the human ear, but as everyone has a different hearing curve, there are a multitude of slightly different gray noise spectra.

This type of noise, like white noise, is also used for the treatment of tinnitus or hyperacusis, that is, increased sensitivity to everyday noises.

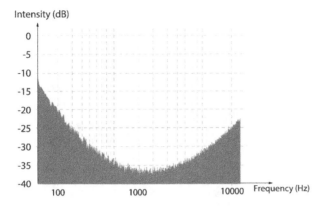

Figure 1.66. *Gray noise spectrum*

1.8.7. *Other noise*

There are other noises, such as green noise, which has a range centered on 500 Hz, orange noise, which is linked to musical scales, creating disagreements, a source of a shattering and quite unpleasant cacophonic noise and finally, black noise, which is in fact the "sound of silence".

NOTE.– Orange noise, typically used for relaxation, is a noise made up of low and medium frequencies located around the zone between 70 and 200 Hz (see the reference section of this book).

1.9. History of sound synthesis

We cannot conclude this chapter without giving a brief history of sound synthesis, which gave birth to many instruments, including synthesizers.

Sound synthesis existed long before the first electric or electronic instruments. This was the case with the church organ that worked using *additive synthesis* (see section 2.2) to generate complex sounds. The instrument is made up of different stops activated by *drawbars* which, when a key is pressed, send air into groups of pipes of different sizes, thus generating distinct timbres such as the drone, the flute, the piccolo and flageolet.

The ancestor of the first synthesizer is most certainly the Dynamophone or Telharmonium, invented and patented in 1897 by Thaddeus Cahill, an American inventor from Iowa. It is a huge machine that weighs several tens of tons and generates sound signals through oscillators made up of alternators driven by electric motors.

Figure 1.67. *An example of church organ pipe groups. For a color version of this figure, see www.iste.co.uk/reveillac/synthesizers1.zip*

The composer Ferruccio Busoni created *Sketch of a New Esthetic of Music* specifically for this instrument.

Figure 1.68. *Thaddeus Cahill's Telharmonium (source: Wikipedia)*

Leon Theremin, whose real name is Lev Sergeyevich Termin, invented the Etherophone in 1920, which was later renamed the theremin (also called the thereminovox). The principle of this instrument is based on two high-frequency oscillators that deliver a fixed frequency and a variable frequency. These two frequencies are inaudible but the hands of the performer, as they move closer or further away from the antennas of each of the two oscillators, cause an audible interference that causes a variation in pitch and intensity. This instrument was used from 1924 by several composers and can be found in many works of contemporary music such as *Ecuatorial* by Edgar Varèse. Today, this instrument is still made and used by many musicians.

Figure 1.69. *(a) Leon Theremin's thereminovox and two modern theremins by Moog, (b) Etherwave and (c) Theremini. For a color version of this figure, see www.iste.co.uk/reveillac/synthesizers1.zip*

The Sphärophon (or Spherophone) appeared in 1926, invented by the German Jorg Mäger. It is one of the first instruments to use triode-type tubes (vacuum tubes) to build its oscillators.

In 1928, the Frenchman Maurice Martenot presented an instrument called ondes Martenot, which used the principle of the theremin. This instrument had a mobile monodic keyboard (see section 3.9) with six octaves, making it possible to carry out vibratos and a ribbon parallel to the keyboard that allowed for glissandi. An expression key manages the sound volume. The diffusion system is very original and is made up of four separate elements, one for the natural sound of the instrument, one for the so-called reverberant long resonance sounds, one for the crystalline and metallic sounds and another for the light and airy sounds; the main "diffuseur" or D1, the resonant "diffuseur" or D2, the gong or D3 and the resonance chamber. Many musical works of all kinds have been and are still written and composed for this instrument.

Figure 1.70. *The ondes Martenot by Maurice Martenot, with four speakers, called "diffuseurs". For a color version of this figure, see www.iste.co.uk/reveillac/synthesizers1.zip*

In 1930, the German Friedrich Adolf Trautwein invented the Trautonium. It is a subtractive synthesis synthesizer (see section 2.1) that uses a low-frequency oscillator. The choice of the pitch of the sound is made by pressing on a wire along a metal bar. It has a very wide range of timbres. Composers like Richard Strauss and Paul Hindemith wrote for the Trautonium.

Figure 1.71. *The Trautonium of Friedrich Adolf Trautwein (source: Technical Museum Vienna). For a color version of this figure, see www.iste.co.uk/reveillac/synthesizers1.zip*

Sound Synthesis 67

In 1934, Laurens Hammond designed an electronic and mechanical organ based on additive synthesis, the Hammond organ, which implemented a system of tone wheels driven by a synchronous motor. A drawbar system composes the timbres. This instrument, designed for classical music, also found its audience among rock and jazz musicians. Its incomparable sound still makes it a widely used instrument today. Hammond organs continue to be manufactured to this day, although they are electro-digital clones of older models.

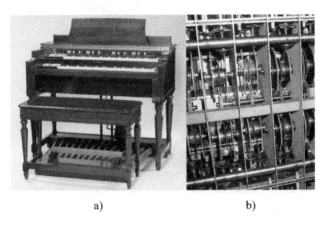

Figure 1.72. *(a) Hammond Model A organ, developed by Laurens Hammond and (b) a photo showing the inner part of the cradle that carries the tonewheels. For a color version of this figure, see www.iste.co.uk/reveillac/synthesizers1.zip*

Figure 1.73. *The ondioline. For a color version of this figure, see www.iste.co.uk/reveillac/synthesizers1.zip*

In 1941, the Frenchman Georges Jenny created the ondioline, a monodic keyboard instrument with three suspended octaves allowing for vibrato and operating with lamps (vacuum tubes). Its selection of timbres allows it to imitate acoustic instruments.

The ondioline has been used in many musical recordings, especially for film soundtracks, such as *Spartacus*, *Baby* and *You're a Rich Man*.

Figure 1.74. *The RCA Mark II[17] synthesizer. For a color version of this figure, see www.iste.co.uk/reveillac/synthesizers1.zip*

Harry Olsen and Herbert Belar developed the RCA Mark I at the Columbia-Princeton Electronic Music Center in New York in 1954, then the RCA Mark II model, known as "Victor" in 1957. These are programmable instruments with several oscillators that generate many waveforms. Data entry is done via two alphanumeric input consoles. Data can be stored on perforated paper tapes. The American composers Otto Luening and Vladimir Ussachevsky (of Russian origin) are pioneers of the project that built these synthesizers.

In the 1960s, electronic organs began to appear, and instruments such as the Continental I and II models (1962 and 1963) from Vox appeared in rock bands. Their technology is based on transistors.

In 1963, a very special instrument appeared on the market. It can be considered the ancestor of our current *samplers* and is not in itself a synthesizer. This is the Mellotron. It was originally designed by the American, Harry Chamberlin, in California at the end of the 1940s. Its name, which comes from "melody electronics" only existed after 1963; when it was created, it was called the Chamberlin MT100.

---

17 Available at: www.synthmuseum.com.

Sound Synthesis 69

Figure 1.75. *The Vox Continental I organ. For a color version of this figure, see www.iste.co.uk/reveillac/synthesizers1.zip*

These instruments were designed in partnership with brothers Frank, Norman and Leslie Bradley, who ran the English company Bradmatics Ltd., located in Birmingham. This company produced mechanical and electronic recording, playback and amplification devices that made up the Mellotron. Many different models were created between 1963 and 1981. The principle remains simple; the sounds are recorded on magnetic tapes placed in interchangeable racks (offering different sound banks) on the most recent models (from the M400 onwards). The early 1980s marked the demise of the Mellotron in favor of modern samplers.

The 1970s and following decades saw the appearance of many synthetic instruments, many of which remain popular today (see Chapter 4). Companies and creators invented and designed synthesizers using the most advanced and popular technologies of their time.

Figure 1.76. *The Mellotron M400. For a color version of this figure, see www.iste.co.uk/reveillac/synthesizers1.zip*

It is impossible to mention them all, but the following: ARP, Moog, Oberheim, Sequential Circuit, Korg, Roland, Fairlight, Yamaha, PPG, Synclavier, Emu Systems and so on, and the geniuses Alan Pearlman, Robert Arthur Moog, Dave Smith, and Donald Buchla, all enriched (and some still enrich) the landscape of sound synthesis instruments. Composers and sound engineers are also important, and it is often under their impetus that new concepts are born.

1.10. Conclusion

After this short overview devoted to sound, acoustics and the history of sound synthesis, you have the basics and the knowledge required to understand the different principles, sound experiments and exercises covered in this book.

We have certainly not reviewed all the aspects of sound, but you can deepen your knowledge by consulting the reference section at the end of the book. However, you should be able to continue reading without any issues.

2

The Different Types of Synthesis

The term sound synthesis means any analog or digital processing of sound signals, which, by means of one or several algorithms and methods, will produce a complex sound.

Sound synthesis is now carried out using two techniques, *analog synthesis* or *digital synthesis*.

In this chapter, we will state the general principles of both analog and digital synthesis.

2.1. Subtractive synthesis

This is the preferred synthesis of analog synthesizers. Its basic principle is based on two main elements, a sound source and modifiers.

The source is a signal that contains harmonics, such as a square, triangular, sawtooth or pulse signal. A sinusoidal signal is not relevant here since it contains no harmonics, it is a pure sound.

The main generator of subtractive analog synthesis is usually an oscillator, whose frequency is controlled by a voltage, called a voltage-controlled oscillator (VCO).

It generally produces continuous sounds that will be modified by processing. Most synthesizers have several of these, which can generate multiple waveforms.

Figure 2.1. *The Novachord, a Hammond product. For a color version of this figure, see www.iste.co.uk/reveillac/synthesizers1.zip*

Many machines use subtractive synthesis.

We can list some of them in chronological order:

– 1939: the Novachord, designed by Laurens Hammond, Polyphonic 12 Top Octave Synthesizer (TOS) oscillators, divider, envelope generator (six shapes available), passive resonator filter;

– 1962: the Synket, designed by Paolo Ketoff and John Eaton, two keyboards each with one VCO type oscillator and a voltage-controlled amplifier (VCF) type filter;

– 1963: the Buchla 100 series, designed by Donald Buchla, modular technology (VCO, VCF, low-frequency oscillator (LFO), voltage-controlled amplifier (VCA), etc.);

– 1965: the Moog Modular, modular technology (VCO, VCF, LFO, sequencer, envelope generator, VCA, etc.);

– 1969: the EMS VCS-3, three VCO type oscillators, one filter, one LFO;

– 1970: the Minimoog (see section 4.1.2);

– 1971: the ARP 2600 (see section 4.1.1);

– 1976: the Oberheim OB-1, two VCO oscillators, one LFO, one filter;

– 1978: the Korg MS-20, two VCO oscillators, two VCF filters, two envelope generators, one noise generator and one wiring matrix.

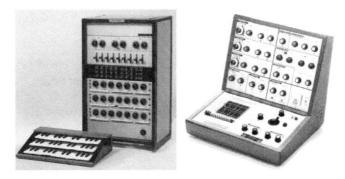

Figure 2.2. *The Synket and the EMS VCS-3. For a color version of this figure, see www.iste.co.uk/reveillac/synthesizers1.zip*

Most VCOs can provide different signals: triangular, square, sawtooth, pulse and so on. In the case of a pulse signal (Pulse Wave Modulation), an LFO is used to vary the width of the pulse.

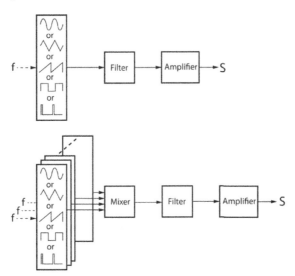

Figure 2.3. *Principle of subtractive synthesis. One or more oscillators (VCO) whose frequency can be controlled are directed to a filter (VCF) and then to an amplifier (VCA). With multiple VCOs, signals are mixed before passing through the filter*

In a subtractive synthesis synthesizer, it is not uncommon to find several VCOs that can be synchronized with one another. Their signals converge on a mixer before joining a modifier, usually a filter to which an envelope is applied.

Behind the filter, there is another modifier, the VCA, which is used to modify the volume of the signal or its shape.

The VCF and VCA modifiers are generally controlled by envelope generators, which transform the evolution of the intensity of the sound over time.

We will come back to each of these elements in detail in Chapter 3.

2.2. Additive synthesis

Additive synthesis adds sinusoidal signals to produce a final sound. This type of synthesis is more complex than subtractive synthesis because many parameters must be controlled simultaneously.

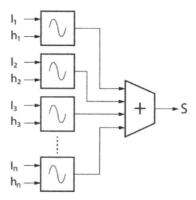

Figure 2.4. *Principle of additive synthesis. A set of oscillators delivers sinusoidal signals whose fundamental h_1 and its harmonics h_2 at h_n with intensity I_1 at I_n are added to form the outputted signal S*

It is based on studies by mathematician Joseph Fourier (see section 1.3.2), who showed that any periodic waveform can be described in terms of the frequency and amplitude of a series of sine waves.

By applying this principle, it is possible to create complex waveforms. The simplest method uses a fundamental frequency and additional frequencies that are multiples of it.

Many simple waveforms, such as square, triangular and sawtooth, are based on this principle; multiple waves are called harmonics.

The church organ is one of the first instruments to work using the principle of additive synthesis, from the end of the 19th century. With the appearance of electrical and electronic systems, others have taken the same path, such as the telharmonium, the Hammond organ (see section 1.9), the Bell Labs Alles Machine and the Kawai K5.

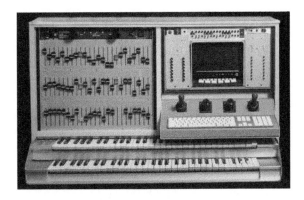

Figure 2.5. *The Bell Labs Alles Machine Additive Synthesizer: right, the control panel for its 72 oscillators. For a color version of this figure, see www.iste.co.uk/reveillac/synthesizers1.zip*

Additive synthesis is not simply based on the sum of several sinusoidal signals but on the harmonic content as well, whose simple phase changes can give rise to major distortions to the waveforms.

Although it seems surprising, the phase change of a harmonic, and therefore waveform, does not necessarily equate to a change in timbre. Very often, the ear does not hear the difference, especially for frequencies above 400 Hz. Thus, phase control can become very useful in certain circumstances to perform additive synthesis.

To obtain interesting timbres close to reality, the user must know the harmonic content of the instruments or sound phenomena they wish to synthesize. This information can be obtained by a Fourier analysis, which will precisely define the harmonic content.

It may seem easy in absolute terms, but it is generally complicated in reality. By scanning the original signal with a narrow band filter, we can successively obtain an intensity in each frequency band of the audible audio range. The narrower the filter, the more accurate the rendering; the bandwidth and sharpness of the filter are therefore very important parameters.

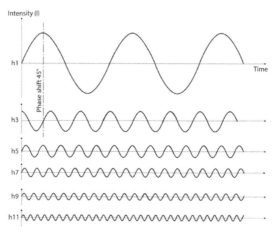

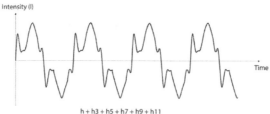

Figure 2.6. *Sum of a fundamental frequency h1 with 5 of its odd harmonics: h3 phase shifted by 45° ($\pi/2$ rd); the shape of the signal obtained is very different to that of a square signal*

With a simple sound, this process is easily applicable, it is quite different with complex sounds where the design of the filters becomes very complicated to obtain convincing results. Over time, with the arrival of digital technologies and the power of computers, most processing has become easier to implement.

Designers of additive synthesizers still ask themselves a question today: how many distinct sinusoidal oscillators are necessary to generate acceptable sounds that are close to reality?

In fact, a good compromise is between 32 and 64, knowing that a C4 at 523.251 Hz, in the middle of the keyboard, will have for harmonic 38 a frequency of 19,883.538 Hz, just at the limit of audible frequencies.

No real sound is composed with only a fundamental frequency and harmonics; it contains many other complementary frequencies such as inharmonic overtones, noise or sidebands.

Inharmonic overtones are additional frequencies that possess a definite shape, without having any relation to the multiples of the signal's fundamental frequency. They are often found in gongs, bells, metallophones or certain other percussive instruments.

Noise, as we saw in Chapter 1 (section 1.8), does not rely on any harmonic structure, it manifests itself as random additional frequencies that appear with indeterminate phases and levels.

When a sound is frequency modulated, or when the frequency of one or more harmonics is fluctuating or imperfect, frequencies are created that are reflected around the frequency considered, causing disturbances in the frequency spectrum of the original signal.

Another phenomenon can also hamper a sound, and these are the beat frequencies, which materialize when the harmonics are not perfectly tuned to each other (see section 1.7.9).

Generating a signal composed of several sinusoidal signals for additive synthesis is not the only challenge. It is also necessary to ensure its envelope, via an envelope generator, is different for each harmonic. All of these envelopes would then be added together to obtain the final envelope. We quickly understand the complexity of such a system and must therefore group several harmonics under the same envelope, such as separate envelopes for the even harmonics, the odd harmonics and the fundamental frequency. You can also choose to work according to the overall pitch of the sound or the note and work by frequency bands.

There is another option, that of using a filter on the overall sound or on groups of harmonics of the overall sound, distributed by frequency bands. The envelope is then applied, similar to subtractive synthesis.

In conclusion, we can say that analog additive synthesis is difficult to implement because it relies on many parameters, frequencies, phases, intensities and envelopes. Although it is possible, it is easier to do digital additive synthesis, where controlling each parameter is easier to achieve.

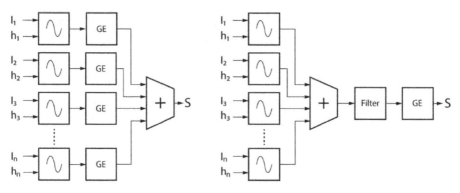

Figure 2.7. *Two of the principles of managing envelopes when creating a sound by additive synthesis*

2.3. FM synthesis

Frequency modulation (FM) synthesis was invented in 1967 by John Chowning at Stanford University. Its principle is based on the frequency modulation of a sinusoidal signal by another signal of lower frequency.

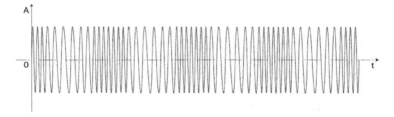

Figure 2.8. *Frequency modulated sinusoidal signal*

The first signal is called the *carrier signal* or *carrier frequency* and the second, the modulator or the *frequency modulation*. When the frequency modulation is lower than 20 Hz, the modulator signal is *infrasonic*, which means that our ear cannot hear it, and we perceive a vibrato effect that alters the sound message of the carrier frequency whose pitch of the sound rises and falls cyclically. Vibrato is therefore a form of frequency modulation.

Beyond 20 Hz, the modulation wave reaches the audible frequencies, and we then perceive a change in timbre.

The synthesis of two sound waves modulating each other produces new sound signals, depending on several parameters such as their frequencies, of course, but also their amplitude and mathematical ratio (frequency modulation/carrier frequency), which can be an integer multiple or not of the fundamental frequency. The sound signal generated can contain several different harmonics or non-harmonics depending on each of the parameters described above. If there are too many non-harmonics, we will no longer distinguish a timbre but rather noise, due to a pitch that is not clearly defined.

FM synthesis is based on simple principles, but its use is not straightforward given the diversity of its parameters.

Also note that sound parameters, such as the presence of sub-harmonics in the output signal, considerably affect the timbre.

FM synthesis is inseparable from the Japanese firm Yamaha. It is the first company to have presented synthesizers using this technology, the GS1 and the GS2, during the summer of 1980 at the NAMM show[1]. Since 1973, it has worked on this project with John Chowning.

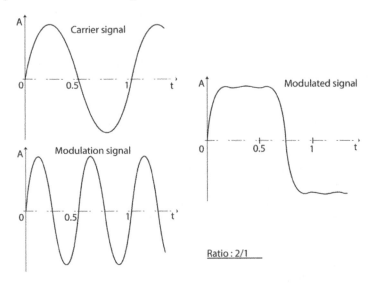

Figure 2.9. *Influence of the carrier frequency/modulation frequency ratio*

---

[1] The NAMM (National Association of Music Merchants) show is an annual event. It is the largest trade fair dedicated to the music sector in the United States.

Yamaha's implementation of FM synthesis involves *operators*. These are modules that integrate a sine wave oscillator, an amplifier and an envelope generator, as shown in Figure 2.10.

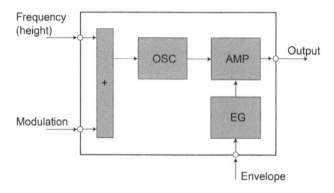

Figure 2.10. *The structure of an operator: in the + block, frequency and modulation data are combined*

An operator has one output and two separate inputs. These are combined and routed to the oscillator.

The first receives the pitch data and controls the internal frequency of the oscillator. The second receives the modulation, which will give rise to FM.

Using several operators, combining them, it will be possible to produce complex sounds. The possible configurations of each of these operators are called *algorithms*.

An operator is said to be a *carrier* operator if its output is directed toward an amplifier to restore the sound signal created. It is said to be a modulator if its output is directed towards the input of another operator. With a carrier operator, the envelope determines the volume of the note produced. With a modulator, the envelope determines the timbre. Carriers and modulators can be joined to form carrier–carrier or modulator–carrier combinations.

As for the algorithms, we will take as an example those available on the Yamaha model DX7[2] synthesizer.

---

[2] The Yamaha DX7 synthesizer was built by Yamaha between 1983 and 1986. It was the first digital synthesizer that conquered the world of music professionals.

The Different Types of Synthesis   81

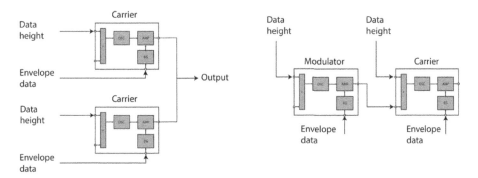

Figure 2.11. *Combination of two operators: carrier–carrier and modulator–carrier combinations*

Figure 2.12. *The DX7 synthesizer from the Japanese firm Yamaha. For a color version of this figure, see www.iste.co.uk/reveillac/synthesizers1.zip*

It has six operators that can be combined according to 32 different algorithms.

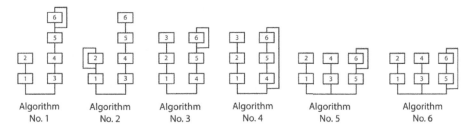

Figure 2.13. *The first six algorithms of the DX7 (six operators are available)*

Operators can be connected in series, parallel or series-parallel. Lower operators are always carrier operators that are connected to the output.

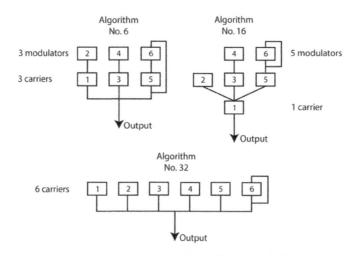

Figure 2.14. *Different types of possible operator connections with their outputs still tied to lower operators*

We will now show how FM synthesis works compared to additive synthesis. Take operators 5 and 6 from Algorithm 4 and apply to them the three sets of parameters in Table 2.1.

Series	Operator	Type	Frequency (Hz)	Intensity (%)
1	5	Carrier	440	100
	6	Modulator	440	100
2	5	Carrier	440	100
	6	Modulator	880	100
3	5	Carrier	440	100
	6	Modulator	1320	100

Table 2.1. *Operator settings*

Here are the waveforms obtained in Figure 2.15.

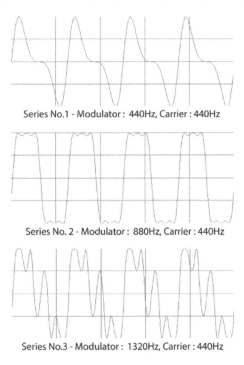

Figure 2.15. *Waveforms obtained by FM synthesis for series 1, 2 and 3*

For additive synthesis, the shapes would have been those of Figure 2.16.

We can see the differences between these two techniques. The waveforms generated are clearly not identical.

FM synthesis also implements envelope data that will modify the transient characteristics of each output signal. As we have seen previously, these data are very important for formalizing the timbre.

DX7 envelope generators are present in each operator. If the operator is a carrier, the envelope simply controls the volume of the output signal. If it is a modulator, the envelope controls the timbre.

The envelope that can be used in the DX7 takes the form shown in Figure 2.17. There are four adjustable intensity levels and ramps (segments representing a rate of weakening or increase of the signal) that connect them.

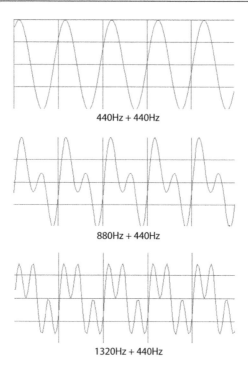

Figure 2.16. *Waveforms obtained by additive synthesis for series 1, 2 and 3*

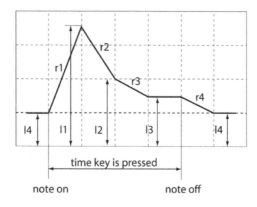

Figure 2.17. *Customizable envelope shape for the DX7*

The *note-on* and *note-off* points correspond to the moments in time when the sound signal begins to be generated by pressing and releasing a key on the keyboard.

Level *l1* is the attack level, which is reached when pressing a key on the keyboard; *r1* is the attack time taken to reach this level. Level *l2* is reached during the decay time *r2*. Level *l3* is the sustain level, which is reached at the time *r3*. The time *r4* represents the release or time taken to return to the initial level *l4*.

The envelope management parameters for the DX7 are more flexible than the classic attack, decay, sustain, release (ADSR), which are managed by many synthesis systems; indeed, the time *r3* integrates a slope to reach level *l3*, the level for which a sound is maintained.

On the DX7, the levels can occur between 0% and 99%, and the ramps are defined by a rate varying between 0% and 99%. A zero rate gives a horizontal line and a rate of 99 a vertical line.

Continuing our discovery of FM synthesis with a concrete example, we are going to synthesize the sound signal generated by an electric piano. For this, we will use algorithm 5 of the DX7.

Operators 1 and 2 will bring the brilliance and metallic resonance close to a bell that nicely characterizes the sound of an electric piano. Operators 3 and 4 will generate the main tone and timbre. Operators 5 and 6 will be used to create the attack sound of the hammer (in electric pianos, a hammer strikes a metal rod acting as a string).

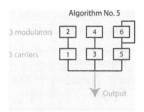

Figure 2.18. *Algorithm 5 of DX7*

Each carrier operator has an exit level and a ratio. The ratio is the number by which the frequency of a note played is multiplied.

For example, for an LA3 (A3) at 440 Hz, if the ratio is equal to 2, the output frequency of the operator will be 880 Hz.

The setting of levels and ratios for the six operators of our electric piano will be that of Figure 2.19. On the DX7, the ratios vary between 0.5 and 61.69 (1,000–9,772 Hz) and the levels between 0 and 99.

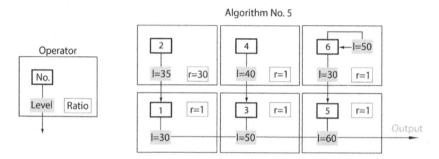

Figure 2.19. *Parameterization of operators to synthesize an electric piano (l represents the level, and r represents the ratio)*

Operator 6 is looped (feedback), which aims to enrich the sound output from the stack with harmonics[3], modulator-porter, consisting of the two operators 6 and 5.

Operator 2 modulates operator 1 with a ratio equal to 30 to generate a sound close to that of a bell.

The waveforms and associated spectra generated by each of the carrier operator pairs for an LA3 (A3) at 440 Hz are shown in Table 2.2.

Operator	Waveform	Spectrum
1, 2		
3, 4		
5, 6		

Table 2.2. *Waveforms*

At output, the signal and the spectrum have the aspect indicated in Table 2.3.

---

[3] When two or more operators are placed on top of each other in an algorithm, we usually call this assembly a stack.

The Different Types of Synthesis 87

Operator	Waveform	Spectrum
1, 2, 3, 4, 5, 6		

Table 2.3. *Signal and spectra*

As for the envelope, Figure 2.20 provides all the elements.

On the DX7, other parameters such as the pitch envelope are available.

They will not be presented here as this would go beyond the scope of this book. Consult the reference section at the end of the book to go further.

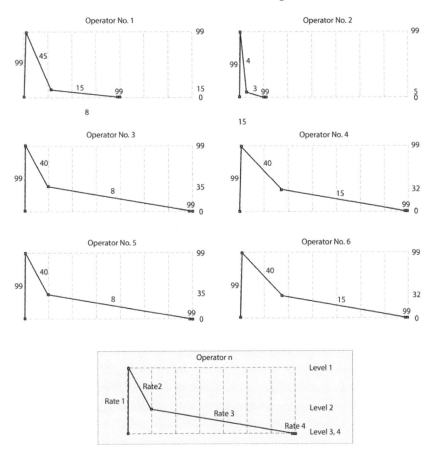

Figure 2.20. *The envelopes of each operator*

2.4. Digital synthesis, sampling and wavetables

Digital synthesis appeared after *FM synthesis*; it uses sounds that were previously stored in memory. To place these sounds in memory, the technique used is called sampling.

Sampling consists of discretizing the signal in order to transform it into a series of values, measured at regular intervals. Periodically, the amplitude of the signal is measured, and the measurement points are plotted. Each point corresponds to a *sample*. Through this technique, a sampling period and therefore a frequency (inverse of the period) are defined. During the reconstruction of the signal, the value of the sample remains constant for the time allocated to the period.

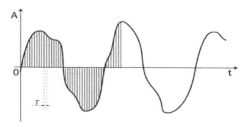

Figure 2.21. *Sampling a continuous signal over the time period T*

During sound synthesis, periodic waves are generally used, that is, repetitive over time. By sampling a period, each of the samples can be stored to memory. The sequential reading of the sequence of these allows for the rebuilding of a period; if we loop the process, we then generate a cyclic audio signal like that created classically by an oscillator.

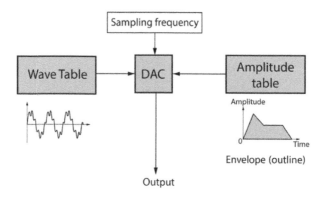

Figure 2.22. *Generation of a sound from a wavetable, at a chosen sampling frequency and following a determined envelope*

All of this stored memory is found in a wavetable. The transformation of the analog sound signal is ensured by an analog to digital converter and the reverse by a digital to analog converter.

By drawing a limited number of slots from the table, for example, one out of two, the period will be twice as short and the frequency twice as high. It is this principle that is applied in wavetable synthesizers to vary the frequency of the signal.

NOTE.– This technique is also used in digital oscillators.

Within a wavetable synthesizer, several tables coexist containing samples of different waveforms, the linking or mixing of one to the other giving rise in this case to more complex sound signals. This process is called dynamic wavetable playback.

When frequencies greater than half of the sampling frequency are present in the sound signal, we are witnessing *spectral aliasing* or *spectral folding*. The frequency of the resulting sample wave, called the *aliasing frequency*, masks the audio information present in the original sound message.

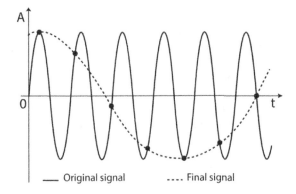

Figure 2.23. *Spectral fallback, with less than two samples per period, reconstruction of the original signal is impossible*

To overcome this problem, it is necessary and sufficient that the original sound signal be limited in its bandwidth and that its sampling frequency is at least equal to twice this band, in direct application of *Shannon's theorem* (Shannon 1949):

> When sampling an audio or other signal, the information conveyed, whose spectrum is bounded support, is completely defined provided

that the sampling frequency is at least twice as large as the largest frequency contained in the signal.

This problem can also be prevented by removing frequencies above half the sampling frequency, before starting the sampling operation itself.

When sound is reproduced via a synthesizer, the wavetable is associated with an amplitude table, which is a curve defining the amplitude (intensity) of the signal over time; it defines the envelope of the sound (see section 3.3).

2.5. Physical modeling synthesis

This synthesis method is based on the simulation of the vibrational processes that produce a sound signal, using a computer. The aim is to describe a sound as an interaction of mechanical and elastic structures. The foundations of synthesis by physical models are based on equations already present in 19th-century scientific treatises, such as *The Theory of Sound* by Rayleigh[4], which discusses the principles of the vibratory systems related to sound waves.

Other physicists and scientists of the 19th century worked on physical models for simulating musical instruments, but it was not until the early 1960s that real applications were created. Researchers like Douglas H. Keefe, James Beauchamp, Pierre Ruiz, Lejaren Hiller, Neville Fletcher and many others are pioneers of synthesis using physical models, applied to the digital processing of music.

In a physical synthesis model, two basic elements are considered, the *exciter* and the *resonator*, which are entry and exit points. Here, therefore, parameters such as mass and elasticity are defined, then conditions defining the limits and the constraints that will be accepted. Envelopes and their transitions (see section 3.3) are also factors that must be considered.

All of these parameters will give rise to a set of complex mathematical equations. These will be processed by the computer, which will provide the amplitude of the sound wave as a function of time, that is, after digital-to-analog conversion, the value of the air pressure at any moment, formalized by the displacement of the membrane of the diffusion loudspeaker(s).

---

4 John William Strutt Rayleigh, better known as Lord Rayleigh, 3rd baron (1842–1919) was an English physicist who published many studies on vibrations and elastic waves. He was also a discoverer (with William Ramsay) of argon gas. He carried out numerous research works related to optics and the theory of light waves and won the Nobel Prize in Physics in 1904.

To model a string instrument, it is likened to a series of masses connected to each other by springs. At rest, this set is in a state of equilibrium. If a force is applied at a point, such as the friction from a bow on a violin string, the masses near that point move relative to their original position.

The disturbances generated then propagate to all the other elements. The compromise between their mass, their density, their elasticity and the resistance of the model defines a propagation speed and a specific release time to the point of equilibrium.

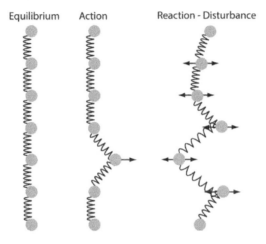

Figure 2.24. *A physical simulation model of a string. An action on one of the masses generates a disturbance in the whole model*

An instrument such as the timpani (also known as a kettledrum) can be simulated by a physical model consisting of a matrix of masses connected by springs.

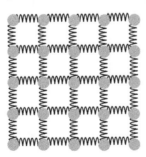

Figure 2.25. *Structure of a physical model for a membrane*

92 Synthesizers and Subtractive Synthesis 1

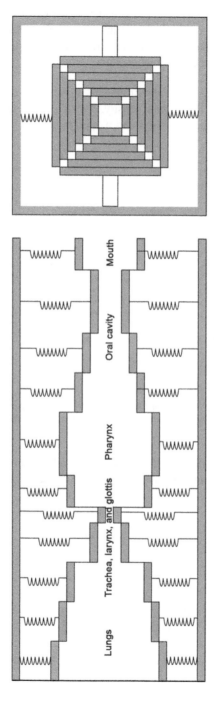

Figure 2.26. *Physical model representing the system involved in the production of speech*

Paul Boersma[5] developed a physical model intended to synthesize the human voice; he was one of the creators of Praat[6], the voice synthesis and analysis software. This software is made up of a set of interlocking square tubes, certain faces of which are associated with springs simulating the muscles present when a human being creates a sound. The tube representing the lungs is closed (diaphragm) and that representing the mouth is open (communication with ambient air).

Physical modeling synthesis is very interesting because all of the elements processed are very significant to the instrument to be modeled, whether real or imaginary. Its limits lie in the monumental quantity of calculations to be processed and in the fact that certain parameters remain very difficult to access, such as the lips or the arm of the instrumentalist.

Figure 2.27. *The Yamaha VP-1, a synthesizer based on physical modeling synthesis. For a color version of this figure, see www.iste.co.uk/reveillac/synthesizers1.zip*

Some manufacturers have made synthesizers using physical modeling synthesis because its approach makes it popular with musicians. However, it remains difficult to operate and master.

2.6. Granular synthesis

Granular synthesis was first proposed by Isaac Beeckman in the late 19th century. However, it was not until 1946 with Dennis Gabor[7] that this synthesis technique gained renewed interest.

---

5 Researcher-professor at IFA (*Institute of Phonetic Sciences*), UVA – University of Amsterdam.
6 Praat is a program for speech analysis and synthesis. It was written by Paul Boersma and David Weenik from the University of Amsterdam.
7 Dennis Gabor, born in Budapest (1900–1979), was a Hungarian-born British physicist. He received the Nobel Prize in Physics in 1971 for the invention of holography.

A *grain* is a sound with a duration varying from 5 to 100 ms. This duration is in fact very close to the perceptible temporal threshold of human hearing.

Granular representations of a sound phenomenon in the form of a cluster of sound grains, known as a sound cloud, are very common in signal processing algorithms and they have been widely processed and used since the 1940s until today. We can cite studies by Gabor, Schroeder and Atal, Helstrom, Harris, Rodet, Crawford, Pierce and so on.

A grain has a starting point, a duration, an envelope and content. Care is taken to ensure that the shape of the envelope is not the source of additional noise or clicks during the grain sequence.

It is of the attack, sustain, release (ASR) type, but it can also be a mathematical function. Envelopes can be stored in a table just like wavetables, but in this case, we refer to envelope tables.

The content of a grain is the waveform it contains, which can be fixed (sine wave, sawtooth wave, etc.), dynamic (like the waveforms produced by FM synthesis) or even natural (grain taken from a sound sample of noise, voice or music).

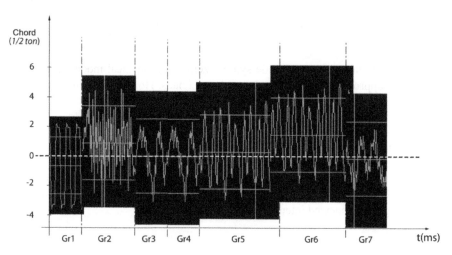

Figure 2.28. *A granular synthesis sequence, seven grains (granules) are assembled at different heights. Gr3 and Gr4 are identical and respect their original height*

The definition of a grain by Dennis Gabor is as follows:

> The grain is a particularly apt and flexible representation for musical sound because it combines time-domain information (starting time, duration, envelope shape, waveform shape) with frequency domain information (the frequency of the waveform within the grain).

Granular synthesis consists of associating several grains to create a sound message, assembled according to different methods by adjusting the height, the positioning, the duration, the repetition, the nesting, the combination, and so on, of several grains (granules).

The grains are read according to a temporal spacing defined by a frequency or a time period. The grains can be spaced out, thus defining a density, which, associated with their length, will make it possible to determine a pitch.

It is also possible to modify the duration without changing the pitch, that is, *time stretching*. The principle is simple: an initial sound is cut into a series of grains, if we want to make it shorter without changing its pitch, selecting a set of grains at regular intervals in this series and bringing them closer is enough. The initial sound is shortened but its pitch remains the same.

This somewhat simplistic application is the basis of time stretching or *pitch shifting*[8] in granular synthesis; it is based on mathematical methods such as the fast Fourier transform and the wavelet transform.

One of the first composers to have applied the techniques of granular synthesis to music is Iannis Xenakis[9] in 1958.

There are many methodologies used within granular synthesis and they generally depend on the organization of the grains, such as Pitch Synchronous Granular Synthesis (PSGS), Quasi Synchronous Granular Synthesis (QSGS), and Asynchronous Granular Synthesis (AGS), among others.

More information can be found in the reference section of this book to complete this brief description of granular synthesis for readers interested in the subject.

---

8 *Pitch shifting* is the opposite of *time stretching*, and the goal is to change the pitch without changing its duration (tempo).
9 Iannis Xenakis, born in 1922 in Bralia, Romania, was a naturalized French composer, architect and civil engineer. Inventor of the concept of "musical mass", "stochastic music" and "symbolic music". Pioneer in the use of the computer for musical composition. President and founder of the *Centre d'Etudes de Mathématique et Automatique Musicales* (CEMAMu), France.

2.7. Amplitude modulation synthesis

Amplitude Modulation (AM) is an old modulation technique that was used extensively in early electronic music.

It is a technique which, like frequency modulation, uses two signals: the carrier and the modulator. The amplitude of the carrier varies according to the modulator.

The difference between FM modulation and AM modulation is that in AM, the modulating signal is unipolar, i.e. always positive.

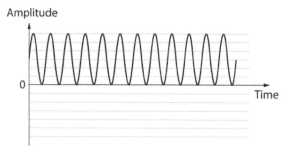

Figure 2.29. *Unipolar waveform, above 0*

Figure 2.29 shows what happens when a sinusoidal signal – the carrier – is modulated by a unipolar envelope type modulating signal. The output signal resulting from the AM modulation, a multiplication of the carrier by the envelope signal, has been shaped.

We also find elements like amplitude modulation when we study ring modulation in Chapter 3 (see section 3.6).

Amplitude modulation generates two sidebands, one for the carrier, the other for the modulating signal. These bands have frequencies equal to the sum and difference of the two signal frequencies. In ring modulation, the frequency of the carrier is not present in the spectrum of the output signal, which is different in AM.

Strictly speaking, there are no AM musical synthesizers, unlike FM synthesizers, for example. However, most analog synthesizers integrate functionalities that allow for amplitude modulation if they have at least two oscillators.

The Different Types of Synthesis 97

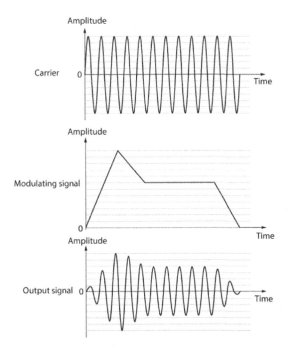

Figure 2.30. *Amplitude modulation of a sinusoidal signal (carrier) by an envelope (modulating signal)*

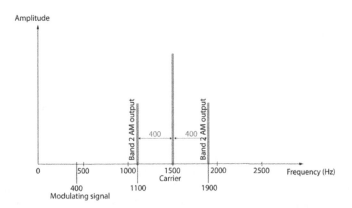

Figure 2.31. *Spectrum of an amplitude modulated signal with its two sidebands and its carrier. Here, the carrier is at 1500 Hz and the modulating signal at 400 Hz*

One of the advantages of amplitude modulation is the ability to generate partially rich signals with only two oscillators. Using a signal with many harmonics, such as a square wave, will create a multitude of sidebands from a minimum number of control parameters. However, controlling the existing partials at the output is complex and delicate, and it is for this reason that AM synthesis is typically used to process signals such as sound effects rather than to create them.

2.8. Phase distortion synthesis

Phase distortion (PD) is similar to FM synthesis. It was engineer Mark Fukuda, from the manufacturer Casio, who developed it for implementation in CZ series synthesizers in 1984. The objective was to manufacture machines that were less expensive than those based on analog subtractive synthesis with filters of the time. The first model marketed was the CZ 101, which appeared in 1985. Other manufacturers have also used this type of synthesis.

Figure 2.32. *The Casio CZ1 synthesizer from 1986, using phase distortion synthesis. For a color version of this figure, see www.iste.co.uk/reveillac/synthesizers1.zip*

PD synthesis provides less coarse sounds than subtractive synthesis, so the idea was to simulate analog filters. Eight waveforms were present from the classic signals, sinusoidal, square, pulse, to more complex signals (see Figure 2.33).

The principle of this type of synthesis is to create the different waveforms by modifying the phase angle of a sinusoidal signal via a second modulation signal, which is similar to FM synthesis. The difference lies in the resynchronization of the signals at each period. To function correctly, both signals, the input signal and the modulating signal, must have an identical frequency. The original sine waveform is stored in memory in a wavetable.

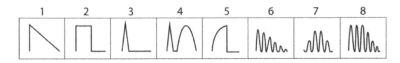

Figure 2.33. *The eight waveforms chosen by Casio*

More simply, we vary, according to a mathematical transfer function, the reading speed of a sine wavetable. The reading range increases from 0 to π (0 to 180°) and then slows down from π to 2π (180–360°). The frequency remains constant, but the shape of the output wave becomes very different from a sinusoidal shape, according to the variation in the phase angle of the modulating function.

By acting on the variables of the transfer function, it is possible to produce an infinity of waveforms and different harmonics from a simple sine wave at the input.

The sounds produced by this technique are often very rich in harmonics.

2.9. Other types of sound synthesis

After presenting the eight most common types of sound synthesis, we are far from having reviewed all the techniques, others do exist but are more rare or have fallen into disuse. Here is a non-exhaustive list, with a brief description:

– *Wave field synthesis*, an extension of wavetable synthesis, which manipulates not tables (in two dimensions) but three-dimensional wave surfaces.

– *Virtual analog synthesis*, which uses the principles of analog synthesis and translates them into virtual synthesis via emulation techniques involving the sampling of signals.

– *Formant synthesis* is based, as its name suggests, on formants, that is, peaks of sound energy composed of harmonics, inharmonics or noise. Vowels articulated by the human voice are good examples, as are the sounds made by many wind instruments.

– *Pulsar synthesis* is based on musical timbres made up of energy pulses, known as *pulsarets*, made up of a waveform followed by a period of silence. It is the repetition of these pulsarets within a pulsar train that forms the final sound. Its time period remains constant but the duration of its waveform, which is modulated, varies, thereby modifying its envelope.

– *Segment synthesis* is a form of digital synthesis that combines samples or fragments of waveforms. It is based on several association and cutting techniques and the interpolation of these forms.

– *Concatenative synthesis* is a synthesis technique that uses sampling, using wavetables and granules, driven by a technology that analyzes, describes and compares the sound. Its main objective is to create very realistic sound textures and atmospheres such as rain, a crowd of people, birdsong, animal cries, atmospheres and so on. Its approach is original since it seeks to build a known sound message by drawing from a library and concatenating very short samples, of around a few tens of milliseconds.

– *Linear arithmetic synthesis* was developed by the manufacturer Roland at the end of the 1980s. It is a skillful fusion of subtractive synthesis and sampled sounds used largely for the sound envelope. The Roland D50 synthesizer incorporates this type of synthesis.

Figure 2.34. *The Roland D50 synthesizer, released in 1987. For a color version of this figure, see www.iste.co.uk/reveillac/synthesizers1.zip*

Over time, new sound synthesis techniques, often derived from existing techniques, appeared.

However, since the 1990s, there has been no real revolution in this area, if we exclude the arrival of digital technology, which has taken over the predicates already present.

Readers wishing to discover other types of sound synthesis, or to deepen those presented here, can consult the reference section at the end of this book.

3

Components, Processing Techniques and Tools

Here, we will see the different components, tools and sound processing techniques present within synthesizers, whether analog or digital.

We will discuss oscillators, filters, envelope generators and amplifiers in this order, to which we will add specific functions such as *sample and hold*, *ring modulation*, *waveshaping*, certain special effects or polyphony.

Finally, we will discuss a more material subject, controllers such as keyboards, dials, selectors and expression pedals.

3.1. Oscillators

An oscillator is a system capable of generating waves in one or more specific shapes, whether sinusoidal, square, triangular, sawtooth and so on.

There are several types of oscillators, including voltage-controlled oscillators (VCOs), digitally controlled oscillators (DCOs) and digital oscillators (DOs).

We also use low-frequency oscillators (LFOs), which are not, strictly speaking, sound generators but signal modulators.

For a color version of all the figures in this chapter, see www.iste.co.uk/reveillac/synthesizers1.zip.

3.1.1. *Voltage-controlled oscillators*

With VCOs, their frequency varies according to a voltage applied to their input. The choice of a control voltage brings flexibility and simplicity to control them.

This voltage can be generated by a keyboard, a sequencer or a MIDI signal converted into voltage (MIDI-CV/Gate, see section 5.1).

The VCOs present in synthesizers have a complex design because they integrate many functions such as tuning, modulation input, range choice or waveform selection.

The input receives the voltage, which determines the frequency of the VCO. This voltage can come from a keyboard, a sequencer or a dial attached to a potentiometer, a pedal or a wheel in order to have progressive variation. A random voltage generator can also be used to obtain random music or even photoelectric or pressure sensors.

The control voltage does not exceed a few volts. The higher it is, the higher the note. The voltage generated by the input device can persist over time.

Depending on the manufacturer, there are different standards, voltage/frequency or volt/octave; we will come back to this in Chapter 5.

The modulation input allows for an adjustable secondary voltage control, which is periodically used to vary the frequency of the VCO to reproduce certain effects such as vibrato or tremolo.

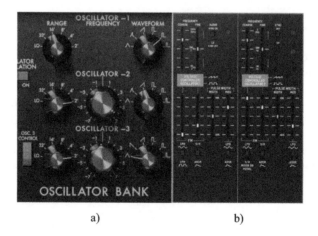

Figure 3.1. *The oscillators of the (a) Minimoog and (b) ARP Odyssey with their different settings: tuning, frequency, waveform and so on*

The choice of range shifts the frequency of the oscillator toward the bass or the treble by one or more octaves. It is generally specified in Hz and kHz or in feet (32', 16', 8', 4', 2', etc.), in reference to pipe organs, 8 feet being the neutral position (the sound played corresponds to the actual pitch of the sound heard). The transition from one range to another automatically transposes all the notes.

A tuning function is often present, it allows for the adjustment of the VCO according to a reference frequency, generally the La3 (A3).

Most oscillators provide overtone-rich sounds so they can be filtered out. Their shapes are usually square, sawtooth, triangular, rectangular and pulse (see section 1.5).

3.1.2. *Digitally controlled oscillators*

One of the major drawbacks of the VCO is its lack of frequency stability and its difficulty in being used for polyphony.

The DCO is an oscillator similar to the VCO but digitally controlled. It has great stability, and the notes are no longer defined via DC voltage but by using numerical values.

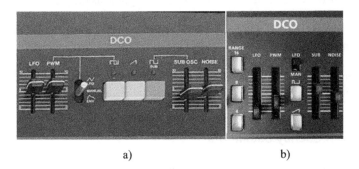

Figure 3.2. *DCO commands of the synthesizer (a) Roland Juno 60 and (b) Roland JU06*

A DCO is a hybrid system that combines analog and digital technology. It marked the entry of synthesizers into the digital electronics era in the early 1980s.

3.1.3. *Digital oscillators*

This type of oscillator is entirely digital. It is used in some synthesizers such as the Clavia Nord Lead or the Access Virus. This type of synthesizer is called a virtual analog synthesizer. These synthesizers can process several forms of sound synthesis and are built around powerful computational processors that simulate analog operation through software components.

a) b)

Figure 3.3. *Two virtual analog synthesizers:
(a) the Access Virus Ti2 and (b) the Clavia Nord Lead A1*

DOs are also the main elements of wavetable s synthesis (see section 2.4).

One of its greatest assets is its stability, but its sound often lacks warmth and remains synthetic.

Another favorite of the DOs is the all-software synthesizer that runs on a microcomputer.

3.1.4. *Low-frequency oscillators*

LFOs produce frequencies generally between 0.1 and 20 Hz. Their purpose is to modulate signals already present to generate special effects.

Like a classic oscillator, it is possible to vary its amplitude and to select its waveform, whether sinusoidal, square, triangular, sawtooth or pulse.

The LFO can be applied to several points of a synthesizer, on the voltage-controlled amplifiers (VCAs), the frequency of the VCO or the cutoff frequency of the voltage-controlled filters (VCFs). It can also act on more specific functions such as panning right to left.

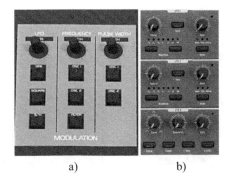

Figure 3.4. *The commands of (a) the Oberheim OB-X LFO and (b) the 3 Arturia PolyBrute LFOs*

The main sound effects that an LFO can create are vibrato and tremolo. It can also contribute to chorus, *flanger* or *phaser* effects (see section 3.8.5).

To create a vibrato effect, the LFO modulates the frequency of the signal delivered by the VCO, while for the tremolo it modulates the volume of the VCA.

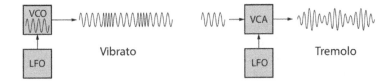

Figure 3.5. *Vibrato: the frequency of the VCO is frequency modulated; tremolo: the volume of the VCA is modulated in intensity*

Similar to a classic oscillator, an LFO can be analog or digital.

3.2. Filters

The filters used are generally VCFs, a voltage is what controls their parameters.

The purpose of the filter is to modify the harmonic content of the sound, more precisely its timbre. There are several types of filters, low-pass, high-pass, band-pass and band-stop (also called rejector or *notch*), which are generally a function of the attenuation curve they provide.

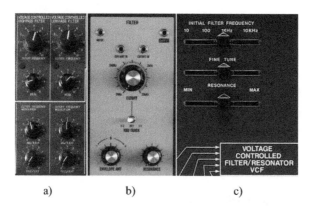

Figure 3.6. *The controls of the filters of the (a) Korg MS-20, (b) Moog GrandMother and (c) ARP 2600*

Each manufacturer uses their own design specification for their filters, resulting in different grains of sound for different synthesizer brands and models.

3.2.1. *Low-pass filters*

This is the most frequently used filter in sound synthesis. The low-pass filter lets the low frequencies pass while attenuating high frequencies, which are determined with respect to a chosen frequency called the filter *cutoff frequency*; the latter separates the two ideal operating modes of the filter, blocking or passing.

The cutoff frequency is the point at which the filter attenuation reaches 3 dB, as shown in Figure 3.7.

Depending on its technology, a filter can have different attenuation curves. There are first- or second-order filters; higher order filters are generally made by combining several filters one after the other, two second-order filters to make a fourth-order filter, for example.

In a synthesizer, we often refer to the number of *pole*s that the filter contains. A *pole* can be considered a pair of RC elements (resistor, capacitor), which are the constituents of a first-order filter.

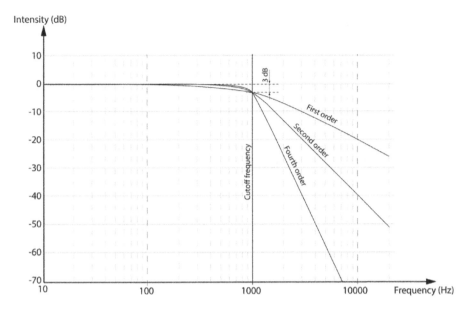

Figure 3.7. Attenuation curve of a first-, second- and fourth-order low-pass filter. The cutoff frequency coincides with an attenuation of 3 dB

A single-pole filter will have an *attenuation* of 6 dB/octave, that is, every time the frequency doubles, the attenuation increases by 6 dB. A two-pole filter will have an attenuation of 12 dB/octave, while a four-pole filter will have an attenuation of 24 dB/octave.

An attenuation of 3 dB corresponds to an input voltage divided by 1.414 (root of 2), which is equivalent to 70.7% of the input signal.

The higher the number of poles, the more synthetic the sound of the timbre when listening. The cutoff frequency is one of the main control parameters for VCFs. As it increases, we first hear the fundamental frequency followed by each of its successive harmonics.

When the filter is set so that only the fundamental frequency is perceived, whatever the shape of the original signal, we obtain an identical sinusoid, which means that when listening the sound remains the same. The harmonics must appear and therefore increase the cutoff frequency with a difference in timbre being heard, depending on the shape of the input signal.

NOTE.– The attenuation rate of a filter after its cutoff frequency is called the *slope*, and it varies according to the order of the filter. During filtering, the phase of the audio signal is modified according to the frequency and therefore the chosen attenuation.

3.2.2. *High-pass filters*

This is a filter that works in the opposite way to the low-pass filter; it attenuates all frequencies below the cutoff frequency.

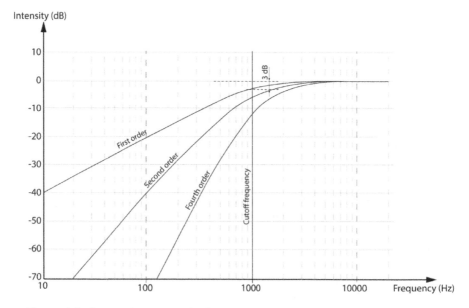

Figure 3.8. *Attenuation curve of a first-, second- and fourth-order high-pass filter*

It begins by eliminating the fundamental frequency and then the harmonics, which can cause the perceived pitch to vary rapidly.

3.2.3. *Band-pass filters*

This is a filter that only passes a selected range of frequencies; all other frequencies are attenuated. The range of frequencies is called the filter *bandwidth* (ω).

On this type of filter, you can adjust the cutoff frequencies and the bandwidth; it can be considered a combination of a low-pass filter and a high-pass filter placed in series.

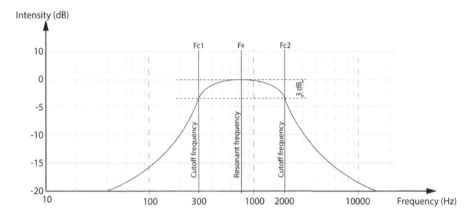

Figure 3.9. *An example of a band-pass filter attenuation curve with its two cutoff frequencies and its resonant frequency*

The central frequency, also called the *resonant frequency* F_R, is equal to the geometric mean of the cutoff frequencies Fc_1 and Fc_2, that is:

$$F_R = \sqrt{Fc_1 \times Fc_2}$$

Taking the elements of Figure 3.9, we have:

$$F_R = \sqrt{300 \times 2\,000} = 774.5 \text{ Hz}$$

The bandwidth ω is equal to:

$$\omega = 2{,}000 - 300 = 1{,}700 \text{ Hz}$$

Band-pass filters can be *narrow band* in which case they can produce significant changes to the timbre of the sound, unlike wider band-pass filters, which accentuate a range of frequencies without making large changes to the sound intonation.

NOTE.– For band-pass filters, there is an important parameter called the quality factor. Q is defined by the equation F_R/ω. Q measures the filter *selectivity*.

3.2.4. Band-stop filters

These are also called notch filters and operate in the opposite way to band-pass filters by defining a range of frequencies to be attenuated over a certain width, limited by two cutoff frequencies.

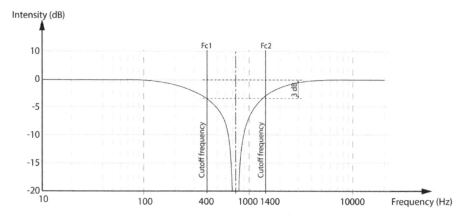

Figure 3.10. *An example of a notch filter curve with its two cutoff frequencies*

The notch filter can be seen as a combination of a high-pass filter and a low-pass filter.

They are often used on narrow bands to suppress a fundamental frequency or one or more harmonics.

Its center frequency is often referred to as the notch frequency.

Like the band-pass filter, we can define its quality via the selectivity Q.

3.2.5. Resonance

Resonance is a peak that accentuates the frequency response of a high-pass filter or a low-pass filter. Typically, resonance is at the cutoff frequency for low-pass or high-pass filters. Most VCFs found in synthesizers use internal feedback to produce it.

Part of the output signal is fed back to the input of the filter; its intensity can be controlled by varying the voltage.

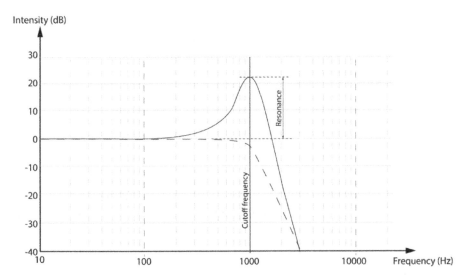

Figure 3.11. *Resonance peak on a low-pass filter*

The resonance brings a richness to the harmonics, which can cause *self-oscillation* of the filter if the gain becomes very large. This self-oscillation produces a pure sine wave.

On band-pass and band-stop filters, resonance becomes a bandwidth control and acts on the selectivity parameter Q. The stronger the resonance, the narrower the frequency band that will be boosted for a band-pass filter or attenuated for a notch filter.

3.2.6. Other filters

Although the filters presented above are the most frequently used in sound synthesis, in particular the low-pass filter, there are other, rarer types of filters:

– Strictly speaking, the *comb filter* is not a filter but a temporal processing of the signal. It adds a delayed version of itself to a signal that causes destructive or constructive interference (see section 1.7.2).

– The *all-pass filter* is a filter that lets all the frequencies pass with an identical gain but modifies the phase between each frequency. This type of filter can be used to create phasing effects (see Volume 2, section 3.3.4).

– The *multimode* or *multipole filter* consists of several dynamically selectable filters. Its application allows an audio signal to evolve by passing it through different filters, thus creating a sound *transmutation* (*morphing*) effect.

Analog filters such as tube amplifiers or tape recordings sound warmer than digital filters. Their technology affects their linearity. They can cause very gradual distortion or saturation, which modifies the reproduced sound. Digital filters can also be subject to this kind of problem, but they are much more intransigent, and the defects often appear unexpectedly, making them more difficult to control.

Each filter has its own personality, which is a function of its technology, internal components and manufacturer. None of them are identical and they never react in exactly the same way, especially analog filters, which are very sensitive to the slightest variations in their environment.

3.3. The envelope generator

The function of the envelope generator is to modulate the amplitude of a sound signal to personalize its contours. It applies to both VCOs and VCFs.

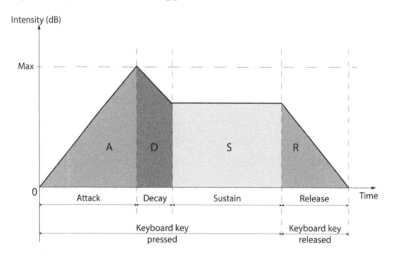

Figure 3.12. *ADSR envelope curve*

The envelope generator acts on the gain of one or more VCA(s). Most synthesizers have two envelope generators, although there may be more on higher-end models.

There are different types of envelopes with different transients, but the most frequently encountered is of the ADSR type.

The envelope generator is activated by what is called the *trigger*, which detects and shapes any external signal, for example, keyboard, dial, sequencer. If the synthesizer has several generators, the trigger will generally trigger them all at the same time.

In special circumstances, the envelope generator can also modify or vary the cutoff frequency of the VCF.

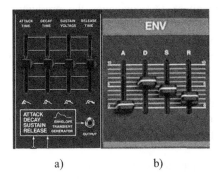

a) b)

Figure 3.13. *Envelope generator controls of (a) an ARP 2600 and (b) a Roland Juno 60*

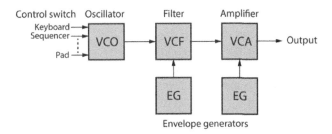

Figure 3.14. *The path of an audio signal*

Within a synthesizer, the usual path of a sound signal has four phases:

– a control device (keyboard, sequencer, pad, etc.) activates a trigger that asks the oscillator to generate a signal;

– the signal is filtered by the VCF;

– its contours can be modified by the envelope generator;

– the VCA amplifies the signal and its contours can be modified by an envelope generator.

Let us take the four parameters presented previously, *attack, decay, sustain*, and *release*.

3.3.1. *Attack*

This parameter, the attack (A), manages the production of the sound. Zero voltage increases until it reaches its maximum intensity, and the attack setting defines the speed at which this increase takes place, that is, its velocity. Generally, the settings provided on the synthesizers vary from a few milliseconds to a few seconds.

Depending on the noise (wind, surf, waterfall, siren, storm, plane, helicopter, etc.) sought or the type of instrument to imitate, the attack is different. It will be very short for percussion instruments and much longer for brass or string instruments.

NOTE.– If the sound is generated using a keyboard or a sequencer, the time the key is pressed on the keyboard or the time assigned to the note on the sequencer must be greater than that of the duration chosen for the attack; otherwise, the maximum level will not be reached.

3.3.2. *Decay*

The decay (D) begins when the maximum attack level is reached. The latter will decrease if the trigger is triggered (key pressed, sequencer note sent, etc.) to reach a value between zero and this maximum level depending on the setting predefined by the user.

The intensity of the sound decreases to reach the level of the next setting, the sustain; however, if the decay setting is very short, the sound produced is very percussive.

In most cases, the decay has an adjustable duration ranging from a few milliseconds to around 10 s, or even more on some digital synthesizers.

Decay is often used to accentuate the attack and give some movement to the audio signal.

3.3.3. Sustain

The sustain (S) keeps the sound at a constant intensity after the decrease brought by the decay and before the extinction of the sound when the control is released (keyboard key, sequencer, etc.).

The sustain can be applied at the end of the attack if the decay is zero. Otherwise, it is applied as soon as the decay reaches a level identical to that defined for the sustain.

In fact, the sustain depends on the duration of the control signal, generally the amount of time a key is pressed on a keyboard. Its adjustment parameter therefore does not act on time but on a voltage equal to or less than the maximum intensity of the signal. Very often, it is expressed as a percentage between 0% and 100%.

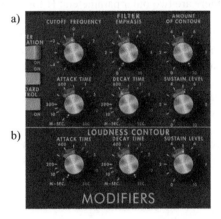

Figure 3.15. *The two Minimoog envelope generators, ADS for (a) VCF and (b) VCA*

When the sustain is close to zero, the envelope sounds more like plucked strings or percussion. With a medium value, it sounds like a piano chord or brassy sounds like the trombone or flute and so on. Long values are geared toward sounds like organs, violins or string ensembles.

3.3.4. Release

The release (R) defines the time that the sound will take to disappear when the control voltage disappears, usually when the user stops pressing the key on the keyboard.

This can provide the link between different notes like the sustain pedal of a piano. The sound is like the remaining resonance after the note or sound has been played, which is common with many musical instruments, vibraphone, cymbal, piano and so on. It can, in some cases, simulate reverb.

The release is adjustable with a typical duration of 2 ms to several seconds.

NOTE.– If the A of the sound is long, for example, 1 s, and the control voltage ceases after half a second, the end of the A, the duration of the D and S will not be considered, and only R will remain.

3.3.5. *Other parameters*

Along with ADSR, there are two other parameters: H for hold and D for delay.

The hold is inserted between the sustain and the release. Its purpose is to extend the duration of the sustain without the sound-controlling device, usually a key on the keyboard, being held down. When the hold ends, the release normally follows.

This parameter is added to some envelope generators to create very long sounds or sound effects such as wind and rain.

The hold does not always appear within the envelope generator but as a function forcing the VCA to let the sound pass continuously. In this case, it is no longer a continuous tuning but an on/off switch.

This delays the trigger of the command, generally the pressing of a key on a keyboard, by up to a few seconds.

Hold and delay allow us to create complex sounds by applying different envelope effects to the fundamental frequency and harmonics of a sound signal.

NOTE.– There are variations of the classic ADSR envelopes, with transient parameters. Attack, delay, sustain (ADS); delay, attack, sustain, release (DASR); or delay, attack, release (DAR) type envelopes can be found on some synthesizer models. The envelopes usually control the VCF and the VCA, but they can also be made to act on other elements of the sound synthesis chain such as effects. There are envelope generators with double outputs, the second of which is said to be complemented, its form being opposite to the first. High-end synthesizers have envelope generators acting on the VCF and the VCA, completely independent, which can be synchronized or cannot be synchronized in their triggering.

3.4. Amplifiers

Amplifiers, also called VCAs, are the final stage of a synthesizer.

They should not be confused with the classic amplifiers that we know for hi-fi or musical instruments. A classic amplifier is a device that transforms an input signal of amplitude A_1 into a signal that has the same shape but is of amplitude A_2. By calculating the ratio A_2/A_1, we obtain a value that represents the gain of the amplifier. A gain of two will provide a signal twice as strong (your ear will not perceive it like this, and the intensity of a signal does not follow a linear scale).

In most audio electronic systems, there is an amplifier. The general volume depends on a volume knob, which controls a preamplifier, whose role is to attenuate the signal input to the terminals of this amplifier. In fact, this knob controls a voltage that ultimately varies the output volume.

This is also what happens in a VCA; however, its voltage can be controlled by an ADSR type envelope generator controlled by a trigger, as shown in Figure 3.16.

The succession of ADSR voltages forms a contour ready to be applied to an audio signal.

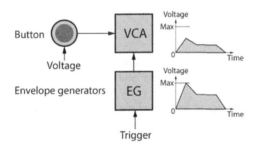

Figure 3.16. *A trigger; for example, pressing a key on a keyboard generates an envelope whose contour is transmitted to the VCA that attenuates it*

The audio signal passes through the VCA, which serves as an amplifier or, more precisely, as a preamplifier by delivering a contour that formats the incoming audio signal, as shown in Figure 3.17.

Instead of being applied to a VCA, the envelope generator can be applied to a filter (VCF) to control the contours of the signal passing through it.

118   Synthesizers and Subtractive Synthesis 1

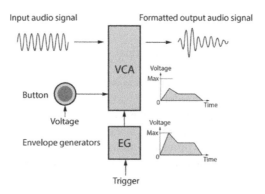

Figure 3.17. *Path of voltage and audio signals through the VCA*

In conclusion, it is important to keep in mind that a VCA amplifies an incoming signal and gives it an outline (a shape). If the VCA is not controlled by an envelope generator, it simply amplifies the incoming signal passing through it.

NOTE.– Like VCOs, there are also DCAs that are digitally controlled. VCAs can be linked, which can be the case within modular synthesizers.

3.5. Sample and hold

As its name suggests, this function samples and then holds a signal until a new one is sampled. It is often abbreviated as S&H or S/H.

To work, the S&H needs at least one input, a trigger signal and an output.

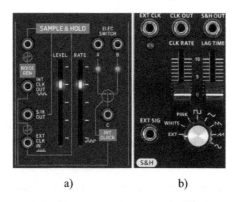

Figure 3.18. *The S&H modules of (a) a Behringer 2600 and (b) a Roland System 500*

When the trigger signal reaches a chosen instantaneous value, a sample of the input voltage corresponding to this value is sampled and then held until the next trigger signal (see Figure 3.19).

A common paradigm is to use an LFO as a trigger, which forces the S&H to sample repeatedly. A common use is to choose white noise as the sampling source, which generates a sequence of completely random voltages, then pass it to an oscillator (VCO), which will play notes that are also random. Instead of the VCO, we can choose a VCF, which will produce a random filtering effect, and the sound will become cacophonous.

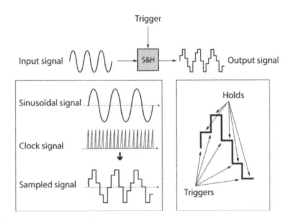

Figure 3.19. *Principle of S&H. A signal, here sinusoidal, enters the S&H, the trigger initiates the sampling of the signal voltage at the required level then keeps it until the next trigger, thereby generating a new sampled signal at the output. The source of the trigger can be a clock signal or a signal from an LFO*

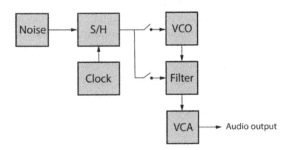

Figure 3.20. *Possible signal routing for the S&H (the clock signal can be provided by an LFO)*

The S&H has a competitor, the T&H or T/H (*Track and Hold*), which works on the falling edges of the clock signal. The sustain is from the top of the pulse and remains constant until the clock signal is again at its maximum. This very different, somewhat reverse, operation considerably modifies the rendering of the outgoing signal, as shown in Figure 3.21.

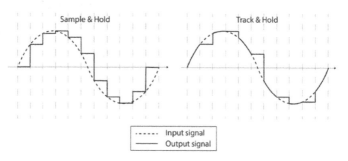

Figure 3.21. *Comparison of S&H and T&H for the same clock signal and the same input signal*

The T&H passes the input signal if the clock signal value is low. When the latter becomes high, it maintains the signal at the detected level.

The S&H signal has a staircase shape, while that of the T&H is much more complex since it consists of alternations of the input signal with its variations and of hold periods.

3.6. Ring modulator

The ring modulator is an audio effect that uses an oscillator to create a sine wave that will be multiplied with an external audio signal. Many synthesizers have one and it is very useful for creating certain metallic sounds like those of a bell.

Its name comes from the configuration of the electronic circuit that formed it when it was designed, a ring of diodes. It was invented by Frank A. Cowan in 1934 for use in telephone multiplexing, to more accurately provide multiple channels to carry multiple telephone conversations over a single cable.

This type of circuit has three connection points, the carrier, the input and the output of the signal. The output of the ring modulator is a mixture of the sum, the difference between the input signal and the carrier, usually a sinusoidal signal.

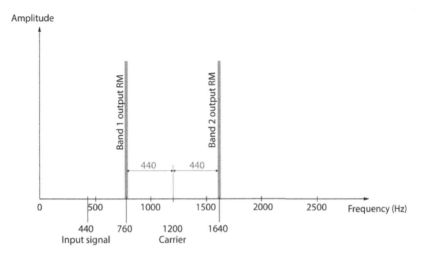

Figure 3.22. *Spectrum of different incoming and outgoing signals of a ring modulator*

Figure 3.22 is an example of what happens with an input signal at 440 Hz (La) and a carrier at 1,200 Hz. The ring modulator results in two signals, one at 760 Hz (1,200 – 440) and one at 1,640 Hz (1,200 + 440).

These two signals form two bands, which are mixed and then directed toward the exit.

NOTE.– If the difference between the input signal and the carrier is negative, for example, a 400-Hz carrier and an incoming signal of 1,000 Hz, that is, 400 – 1,000 = –600, a sideband occurs at the unsigned frequency 600 Hz, and this is true for any negative frequency.

For a very metallic effect, the frequency of the carrier must be two to three times higher than the input signal.

NOTE.– Ring modulation is often confused with amplitude modulation because they are very similar to one other. In both cases, it is a question of multiplying a signal with another simpler signal. If this signal (e.g. a sinusoidal one) is bipolar (positive and negative), ring modulation is created. If the modulating signal is unipolar (positive only), it is amplitude modulation.

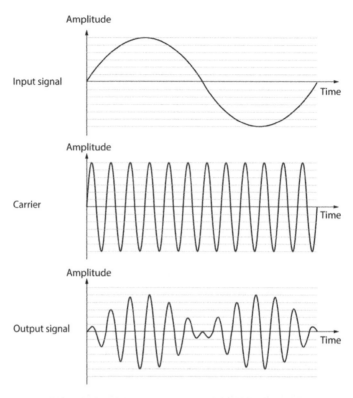

Figure 3.23. *Ring modulation with an equal carrier at 12 times the frequency of the incoming sine wave. You can see the variation in amplitude generated on the outgoing signal*

3.7. Waveshaping

Considering the previous explanations, it seems that subtractive sound synthesis is based on techniques for modifying and enriching waveforms to provide them with the timbre, dynamics and height that will give the desired sound.

Waveshaping is a family of sound wave modification techniques that transform waves to manipulate their harmonic content. This process adds harmonics to or removes harmonics from a complex waveform or, conversely, adds harmonics to a simple wave. To understand the principle of waveshaping, we must begin by explaining the transfer function concept. It is a mathematical function that transforms input values to new output values, as shown in Figure 3.24.

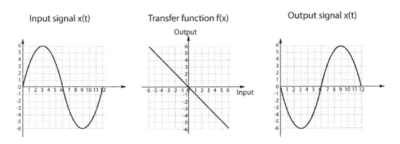

Figure 3.24. Application of the transfer function f(x) on a sinusoidal signal x(t)

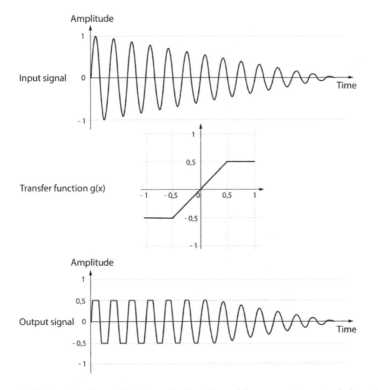

Figure 3.25. Application of the transfer function g(x) on an attenuated sinusoidal signal. The output signal is clipped as soon as its amplitude exceeds ±0.5

In this case, the horizontal axis represents an input signal and the vertical axis an output. If we plot the transfer function $f(x)$ (a diagonal passing through the origin from the upper left corner to the lower right corner $f(x) = -x$), through the latter, the outgoing signal will be an inverted version of the original.

By applying this same principle on more complex transfer functions, we can convert an incoming waveform into any other output form.

Waveshaping includes different shaping methods such as *wavefolding*, *wavewrapping* and *clipping*.

Figure 3.25 shows another example on an attenuated sinusoid with the application of the transfer function $g(x)$.

3.8. Special effects

Very often, effects are associated with sound synthesis and enrich the signal by playing on the sound rendering. We have already encountered two of them, vibrato and tremolo, often generated by an LFO (see section 3.1.4), but there are others such as *pitchbend, glide, keyboard tracking, phaser, chorus, flanger, delay, reverb* and so on. While some of them are integrated into the internal functioning of the synthesizer, others act on the output signal and may be external devices.

3.8.1. *Pitchbend*

This is an effect generally controlled by a knob, a strip or a touchpad. Its objective is to give the user the option of modifying, in real time, the pitch of the sound up or down within a defined range.

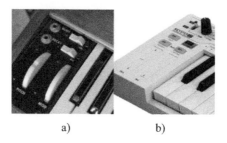

a)   b)

Figure 3.26. *Pitchbend wheel and strip of (a) a Minimoog and (b) an Arturia Keystep keyboard*

As a MIDI message, the pitchbend has the particularity of operating on 16,384 levels. In fact, it is coded in 2 bytes of 7 data bits, that is, 14 bits with the middle position of the wheel defined at the value 8,192. It is for this reason that most software displays pitchbend ranges between –8,192 and +8,192.

The variation retained by the GM standard (see section 5.2.9) is ±2 semitones, but some controllers have a larger scale that can go up to ±12 semitones.

Depending on the width of the variation, the height is stretched respecting the 16,384 levels in all cases.

3.8.2. *Glide*

The glide, also called *portamento*, reproduces all the frequencies located between two notes by performing a continuous glide from one note to another.

The setting associated with this type of effect varies the time it takes to perform the glide. A small value favors a fast sliding and a large value a slower sliding.

On some synthesizers, there are two adjustment modes, normal and slide. In the first, the portamento is triggered only if the notes played are linked together, while in the second, the effect is always active.

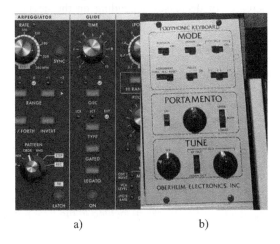

Figure 3.27. *The portamento setting on (a) an Oberhein FVS-1 and (b) a Moog Sub37*

There may also be a legato mode, which activates portamento only when you press a key while holding down a previous key (see also section 3.9).

Other features are sometimes present, and these are as follows:

– The *linear constant rate* setting for which the glide rate depends on the size of the interval between the two notes played. The wider it is, the longer the slide.

– The *linear constant time* setting for which the sliding time is always constant, regardless of the spacing of the interval.

NOTE.– Depending on the synthesizer manufacturer, there are other less common options and adjustment modes for portamento, different from those mentioned above.

3.8.3. *Keyboard tracking*

This function, also called *key tracking* or *key track*, allows you to adjust the cutoff frequency of a filter according to the frequency of the note played. If it is not applied, the cut will only act around the chosen note for which the filter has been set; as soon as we deviate from several notes, with the frequency having changed, the filter will no longer do its job and it will then have to be reconfigured.

Depending on the type of synthesizer, keyboard tracking can take several forms, and it can be centered on a pivot note, usually Do3 (C3), around which the filter cutoff frequency increases or decreases depending on the deviation relative to the pivot note.

Figure 3.28. *The Arturia MiniBrute synthesizer with its keyboard tracking setting (KBD Tracking) located in the filter section*

On other machines, it can be positive and negative, thus changing the distribution according to the choice of a note played lower or higher.

A percentage setting, with values greater than 100%, is present on some synthesizers; it can reach 200%, or even more, modifying the action of the filter proportionally.

NOTE.– Key tracking applies primarily to the filter, but it can also track volume by increasing it on high notes and vice versa. On some digital synthesizers using sampling, key tracking is called *key follow* in order to modify the pitch of a sample according to a key chosen on the keyboard.

3.8.4. Reverb and delay

Many synthesizers incorporate a digital or analog (spring) reverb effect, and this is the case, for example, of the ARP 2600.

Reverb or delay is part of the family of time and spatialization-based effects. They exist naturally, in the mountains, a tunnel, a church, and so on. They create a feeling of space, adding presence to the sound reproduction.

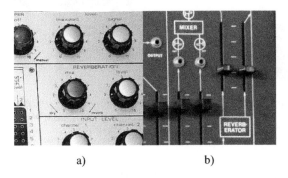

Figure 3.29. *The reverb settings on (a) an EMS Synthi A and (b) an ARP 2600*

It is necessary to distinguish reverb from echo, which is repetition of the sound linked to a very long propagation and reflection time.

Without going into too many details, reverb has many parameters:

– the type, which defines the depth (*room, plate, hall*, etc.);

– the size or density, which defines the depth or density;

– the decay, which represents the decline over time;

– the mix, source/effect, which manages the balance where the sound and the effect enter (dry, wet, etc.);

– the *pre-delay*, which anticipates the action time of the reverb in relation to the sound.

On most synthesizers, these settings are very limited, which is not the case if you use external devices such as specialized modules for modular synthesizers, racks or effects pedals.

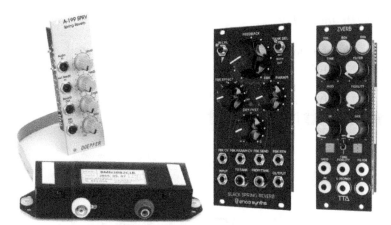

Figure 3.30. *Three reverb modules for modular synthesizers (Doepfer A-199 SPRV, Erica Synths Dual FX and Tiptop Audio ZVerb)*

The delay leans more toward an echo phenomenon (echo chamber, analog or digital delay) and its parameters are different. We will find the duration of the delay, the *feedback* that defines the number of repetitions and so on.

The delay is rarely integrated into synthesizers and uses external devices, but it can be very useful to simulate certain sounds and sound environments.

3.8.5. *Phaser, chorus and flanger*

These three effects are part of the modulation effects[1], and only a few synthesizers have them originally (Arturia MatrixBrute, Juno-60, JX-8P, etc.). This is more often what racks or external pedals are used for. Their purpose is to modify the thickness of the sound in order to obtain a richer sound reproduction.

Both the flanger and the phaser are based on the total or partial reinjection of the original signal from its output to its input, with little delay, which, after mixing, brings a phase shift between the two signals. Some parts of the overall signal are diminished, and others increased, due to interference (see section 1.7.2).

---

1 See Réveillac (2017).

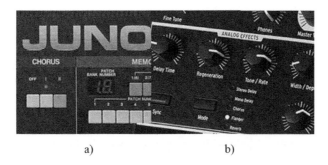

Figure 3.31. *The modulation effects on (a) a Roland Juno 60 and (b) an Arturia MatrixBrute*

Chorus, as its name suggests, provides a chorus effect by multiplying the number of voices with slight differences between them. These are obtained by a delay, a slight asynchronous vibrato or even a subtle modification of the fundamental frequency of the sound.

3.9. From monody to polyphony

Synthesizers were *monodic instruments*[2] until the mid-1970s, capable of playing a single note at a time, then, as technology evolved, they became polyphonic, possessing several voices.

The arrival of digital has dramatically increased polyphony, the number of voices being much less limited, with up to 64, or even 128 voices for a software synthesizer.

The number of voices does not always correspond to the number of oscillators, and most monodic synthesizers have several. Nor is it always equivalent to different keys producing different notes. A single key on the keyboard can produce a multi-note chord.

Monody is based on a single note played at a given time, but what happens when the user presses several keys at the same time? In fact, there is a notion of priority to determine which note should be kept. This priority aspect is different depending on the synthesizer manufacturers. In the 1970s, the Americans gave priority to the lowest note and the Japanese to the highest note, which can exert an influence on the

---

2 It is common to use the term monophony for a single voice; however, the correct terminology is monody, although many consider them to be synonymous. Monophony is more about broadcasting or listening to audio, as opposed to stereophony.

musician's playing and their connection, and the melodic transition of the notes between them.

In parallel with this priority, there is the notion of legato mode, which plays on the triggering of the envelope and the LFO at each note. When this mode is activated, these do not re-trigger, which allows for a smooth transition between each note.

NOTE.– Do not confuse legato with portamento, also called glide, which performs a frequency shift to move from one note to another.

Polyphony is the ability to play multiple voices, which implies that each has its own production chain, oscillator, envelope, filter and amplifier. It is easy to imagine the complexity of a fully polyphonic system, so the creators decided to limit the polyphony to a certain number of voices.

Figure 3.32. *The Polymoog Keyboard, one of the first polyphonic synthesizers, dating from 1975. It had 71 polyphonic notes, a record at the time, and was extremely complicated and expensive*

Many synthesizers are therefore limited to a few voices, the Sequential Circuits Prophet 5 has five voices, the Behringer Deepmind 6 has six voices, the Korg Prologue and the Yamaha DX7 have 16 voices and so on.

Monody and polyphony are not the only criteria expressing the number of voices of a synthesizer, and the term *paraphony* is another system often in use today, as on the Behringer Neutron, the Waldorf Pulse 2 or the Arturia Matrix Brute. This is a mode that allows you to simulate polyphony from a monodic system.

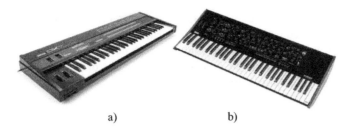

Figure 3.33. *Two polyphonic synthesizers: (a) the middle Yamaha DX7 of the 1980s and (b) the Korg Prologue, a current synthesizer*

The explanation is simple: on a monodic synthesizer with several oscillators, each oscillator can be assigned a trigger on a different note while keeping the same sound processing chain (filters, envelopes, LFO, etc.).

Figure 3.34. *Two paraphonic synthesizers: (a) Behringer Neutron and (b) Waldorf Pulse 2*

To conclude this section, it seems important to present the notion of *multitimbrality*, often confused with polyphony.

For a synthesizer, multitimbrality is the ability to produce sound signals of different natures in terms of timbre, envelope and amplification. They can be distributed at different locations on the keyboard – this operation is called *split keyboard* (e.g. a bass for the left hand and a piano for the right hand) – or even be triggered simultaneously.

In the 1970s and 1980s, musicians had almost as many synthesizers as they wanted to produce different sounds; today most machines are multitimbral, apart from a few monodic analog synthesizers.

Recent synthesizers use patches to store combinations of different layered or distributed sounds.

NOTE.– The patches can be superimposed in a so-called *layering* mode (several layers). Layering reduces the number of available voices. It is also possible to duplicate or even multiply the same voice for each note played, the resulting sound is often thicker, especially if the voices are slightly out of tune (*detune*). This last technique is defined as *unison* mode.

3.10. Controllers

They can be internal or external, they are keyboards, wheels, pads, pedals, joysticks, breath controllers and other more original ones like the keytar, the laser harp and so on.

Whether monodic or polyphonic, all synthesizers use controllers. Although the keyboard is the most common, others are often necessary to obtain more advanced features or better ergonomics.

Most communicate with the synthesizer with CV, MIDI, USB-MIDI or proprietary protocols or interfaces.

NOTE.– Newer MIDI keyboards are often able to handle *velocity* and *aftertouch*. Velocity corresponds to the speed of depression of a key and aftertouch plays on the sensitivity after the depression of a key (control of the pressure exerted).

3.10.1. *Modwheel*

The modwheel, or modulation wheel, is generally placed next to the pitchbend control (see section 3.8.1), often also consisting of a wheel. Compared to the latter, which has a spring to return it to its central position, the modulation wheel retains its given position; it is generally used to adjust the intensity of the LFO but can be assigned, as on recent synthesizers, to any MIDI parameter.

NOTE.– The modwheel corresponds to MIDI CC (Control Change) No. 1 (see section 5.2.6). Instead of a wheel, you can find sensitive bands (see Figure 3.26).

On some synthesizers, the pitchbend and the modwheel are present in the form of a *joystick*, the action of the pitchbend being vertical and that of the modwheel being horizontal.

Figure 3.35. *The joystick on the Yamaha Genos*

3.10.2. *Breath controller*

The *breath controller* is a device that detects the strength of breath. Generally, it modulates the volume or the action of a filter.

This type of controller has been around since the early 1980s. Yamaha marketed it for its CS01 and DX7 synthesizers. Its popularity came from the fact that it became possible to control a synthesizer like a wind instrument.

Figure 3.36. *The Yamaha BC-1 breath controller*

3.10.3. *Expression switch and pedal*

Selectors are simple pedals equipped with one or more switches. They activate a function like sustain on a synthesizer (equivalent to the damper pedal on a piano).

NOTE.– There are so-called "half-pedal" selectors producing an intermediate effect, simulating the light pressure of the sustain pedal on a piano.

In addition to selectors, there are more sophisticated pedals called expression pedals, which are generally used to adjust the volume, the action of the filter or the effects in order to add expressiveness to the user's playing.

Figure 3.37. *Expression switches and pedals (Yamaha FC4A; Roland DP10; Yamaha FC7; Doepfer FP5)*

3.10.4. Keytar

The keytar (from keyboard and guitar) is the equivalent of a guitar whose strings and neck have been replaced by a keyboard. Many manufacturers have built keytars, including Alesis, Roland, Korg, Moog and Yamaha.

In fact, today, the keytar is not always classed as an instrument but as a control device that works via the MIDI protocol, like the external keyboard of a synthesizer. However, there are still keytars that are real synthesizers (Roland AX-Edge, Korg RK-100S, etc.).

NOTE.– The first keytars were real instruments, often electronic organs or synthesizers like the famous Moog Liberation (1980), Korg RK-100 (1984), Yamaha SHS-10 (1987) or the Casio AZ-1 (1986).

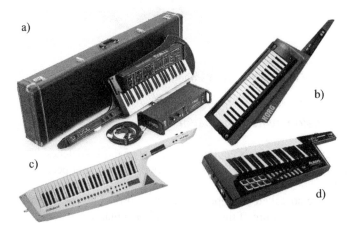

Figure 3.38. *A selection of keytars: (a) Moog Liberatio, (b) Korg RK-100S, (c) Roland AX-Edge and (d) Alesis Vortex Wireless 2*

With the advent of Bluetooth technology, many MIDI keytars have become wireless devices. They are generally connected to synthesizers, and their control device can trigger various MIDI messages.

3.10.5. *Other controllers*

In terms of controllers, everything is imaginable, from the famous laser harp used by Jean-Michel Jarre to super sensitive keyboards like the Roli Seaboard, the Neova ring from Enhancia and many others.

However, all of these non-standard controllers generally work using MIDI CCs, that is, the control codes (see section 5.2.6) specified in their implementation chart (see section 5.2.8).

There are some exceptions, especially for vintage hardware or for some controllers dedicated to modular synthesizers, which can work using CV/Gate.

4

Work Environment

Before discussing our experiments on subtractive synthesis, I will present the different hardware and software used. It goes without saying that if you do not have any, using other resources is possible and should only require minor modifications. On the other hand, this choice has given way to several pieces of software that can be downloaded for free, which should satisfy most readers.

4.1. Materials

Let us start with the materials. We will see the ARP 2600, Minimoog, Behringer Neutron, Novation Bass Station II and Arturia MatrixBrute. The first two are vintage synthesizers from the 1970s–1980s, which over time have become benchmarks. Their popularity is so great that several manufacturers have designed clones very close to the originals. The others are newcomers in comparison, 2013 for the Bass Station II, 2018 for the Neutron and 2016 for the MatrixBrute.

These choices are not by chance, they are all synthesizers working in subtractive synthesis. They cover, because of their characteristics and their architecture, most of the possible configurations in this field, the most recent with features specific to current instruments.

NOTE.– For some machines, especially the most recent ones, which are often complex, their presentation will be followed by a simple description of their operation so that the reader can better understand the structure and architecture of their internal modules.

For a color version of all the figures in this chapter, see www.iste.co.uk/reveillac/synthesizers1.zip.

4.1.1. *ARP 2600*

Alan Pearlman, an excellent engineer passionate about sound from an early age, was the founder of this instrument. In 1948, when he was a student at the Polytechnical Institute of Worcester, he wrote, in an article devoted to electronic music:

> The electronic instrument's value is chiefly as a novelty. With greater attention on the part of the engineer to the needs of the musician, the day may not be too remote when the electronic instrument may take its place as a versatile, powerful, and expressive instrument.

You could say he was a visionary. A little later, in the late 1960s, he heard a version of *Switched on Bach*, recorded and created by Wendy Carlos. The interpretation of several key compositions by Johann Sebastian Bach on a Moog synthesizer led him to design his own instrument, which he called the ARP, his initials.

Figure 4.1. *The ARP 2600 and one of its clones, the Behringer B2600*

The first model to be marketed bore the name ARP 2500. Compared to the Moog synthesizer, which reigned supreme at the time, it brought a great novelty: the user interface was much simpler to use, and there was no need for cables to connect many inputs-outputs to connect each between sound modules, and a simple linear slider was enough. On the other hand, the electronics were much more stable in terms of the oscillators, the centerpiece that produces the signals necessary for the sound synthesis, as their frequencies do not vary as the temperature of the various components rises.

Very few copies (about a hundred pieces) of the ARP 2500 were produced between 1970 and 1981.

Its successor, the ARP 2600, much less imposing, was built around pre-wired modules, which offered less customization but greater ergonomics and greater ease in creating sounds. Its basic architecture is not modular like the 2500 but semi-modular, built around three oscillators, a 24 dB/octave filter, an attack, decay, sustain, release (ADSR) envelope, a voltage-controlled amplifier (VCA), a mixing section, an envelope follower, a ring modulator, a noise generator, a low-frequency oscillator (LFO) and a spring reverb. A stereo amplifier and two speakers complete the assembly.

Figure 4.2. *ARP 2500*

By using pre-wiring, sliders and switches, it is possible to create a wide variety of sounds. However, the user can also build their own signal routing by wiring the different modules to give free rein to even more inventive sounds.

During its period of manufacture, the ARP 2600 underwent some changes, giving way to four generations of different machines:

– The 2600 model, produced in early 1971, known as the "Blue Marvin", was integrated into a wooden and blue aluminum box. Only 25 pieces were made more or less by hand.

Figure 4.3. *The ARP 2600 "Blue Marvin"*

– The 2600C model, produced in mid-1971, with a gray front end, was called the "Gray Meanie". Thirty-five models were made.

Figure 4.4. *The ARP 2600 "Gray Meanie"*

– The 2600P v1.0 was more widespread and integrated into a suitcase-style case to facilitate its transport. Its electronics were identical to the previous model, but it had some reliability issues.

Figure 4.5. *The ARP 2600P with its suitcase-style case*

– The 2600P v2.0 model, produced in 1972, was a version in which some electronic components had been replaced, making the machine much more reliable.

– The 2600P v3.0 model, produced from late 1972 to 1974, was the most common model, again with some internal changes to the electronics. Note that the ARP logo on this model was modified and a treble clef was added.

– The 2600P v4.0 model, produced in mid-1974, came with a new duophonic keyboard and an LFO module. Here again, the electronics had evolved, and certain components were replaced, in particular those of the filter.

– The 2601 v1.0 model, produced from 1975 to 1976, was identical to the 2600P model with a few minor modifications, particularly to the connectors.

– The 2601 v2.0 model, produced in 1977, had a new filter and a new front side, printed in orange on a black background.

– The 2601 v3.0 model, produced at the end of 1980, was housed in a different case. This is the latest model made by ARP.

Figure 4.6. *The ARP 2601*

In January 2020, the ARP 2600 was reborn from the ashes thanks to the Japanese company Korg and David Friend, who co-founded ARP Instruments in the 1980s with Alan Robert Pearlman.

This new model, called ARP 2600FS (Full Size), had the necessary improvements to bring it up to the level of current analog synthesizers: a USB MIDI port, a MIDI In-Out-Thru port, balanced XLR audio outputs, a two-voice (duophonic) keyboard with aftertouch, an LFO and an arpeggiator-sequencer as well as two types of filters.

Figure 4.7. *Korg ARP 2600FS*

In July 2021, Korg presented a smaller version (40% smaller than the FS), the ARP 2600M (Mini) model with new functionalities, in particular MIDI compatibility to adapt to any type of "class compliant" controller; pitchbend, modulation and portamento support MIDI CC codes, XLR outputs gave way to 6.35-mm jacks and a spring reverb is present.

Figure 4.8. *Korg ARP 2600M*

As the ARP 2600 was an exceptional machine, Korg was not the only manufacturer to clone it, and other manufacturers have also entered the race. We can cite synthCube, which markets a model in kit form for self-assembly or completely assembled and checked: the TTSH (Two Thousand Six Hundred).

Figure 4.9. *The synthCube TTSH version 4*

Behringer, another key player and manufacturer of ARP 2600 clones, released the B2600 in October 2020 at a very reasonable price while remaining very faithful to the original, even though the ergonomics are somewhat different. Two types of filters are present, creating different versions, with an LFO, a duophonic mode, a MIDI IN and THRU port and a USB MIDI port.

On the other hand, no amplifier or speakers, just a stereo output via two 6.35 mm jack connectors. The machine is integrated into a metal box with the dimensions of a standard rack.

Figure 4.10. *The Behringer B2600*

Behringer provides three versions of its B2600, the base model featuring digital reverb, the "Blue Marvin" and the "Gray Meanie" incorporating analog spring reverb and premium electronics.

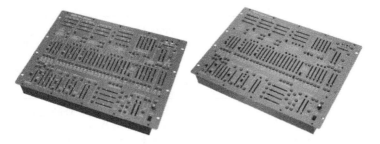

Figure 4.11. *The B2600 "Blue Marvin" and "Gray Meanie" models*

Table 4.1 presents the main characteristics of the different ARP 2600 models according to the manufacturers.

Manufacturer	Model	Year	Oscillators	Filter	Comments
ARP Instruments	2600	1971	3 oscillators (4011)	1 filter (4012)	Blue color – metal box Clavier 3604
	2600C	1971	3 oscillators (4011)	1 filter (4012)	Gray color – metal box Clavier 3604C
	2600P v1.0	1971	3 oscillators (4017)	1 filter (4012)	Dark gray color in suitcase-style case 1971–1972 Clavier 3604P
	2600P v2.0	1971–1972	3 oscillators (4027)	1 filter (4012)	Dark gray color in suitcase-style case
	2600P v3.0	1972–1974	3 oscillators (4027)	1 filter (4012)	Dark gray color in suitcase-style case
	2600P v4.0	1974	3 oscillators (4027)	1 filter (4012)	Dark gray color in suitcase-style case Duophonic keyboard 3620
	2601 v1.0	1975–1976	3 oscillators (4027)	1 filter (4012)	Dark gray color in suitcase-style case
	2601 v2.0	1977–1979	3 oscillators (4027)	1 filter (4072)	Dark gray color in suitcase-style case
	2601 v3.0	1980	3 oscillators (4027)	1 filter (4072)	Dark gray color in suitcase-style case
Korg	2600FS	2020	3 oscillators	2 filters of your choice	Dark gray color Duophonic keyboard 3620
	2600M	2021	3 oscillators	2 filters (4012 or 4072)	Dark gray color Clavier MIDI or Class Compliant Controller
synthCube	TTSH	2016 (v3) 2019 (v4)	3 oscillators	2 filters (4012 or 4072 then v3 Rev 8)	Self-assembly kit version or buy assembled and tested Version 3 or 4

Manufacturer	Model	Year	Oscillators	Filter	Comments
Behringer	B2600	2020	3 oscillators	2 filters	Dark gray color, orange front detail – metal case rack format Digital reverb
	B2600 Blue Marvin	2020	3 oscillators	2 filters (4012 or 4072)	Blue color – metal case rack format Spring reverb Top quality components (op amps)
	B2600 Gray Meanie	2020	3 oscillators	2 filters (4012 or 4072)	Gray color – metal case rack format Spring reverb Top quality components (op amps)

Table 4.1. *Characteristics of the different ARP 2600 models*

In summary, here are the characteristics common to all ARP 2600s:

– monophonic – analog;

– three oscillators (voltage-controlled oscillator, VCO);

– one filter;

– one noise generator;

– two envelope generators (ADSR – AR);

– one reverb;

– one ring modulator;

– one sample and hold function;

– one mixer;

– one headphone jack;

– CV/Gate connections.

Figure 4.12 shows the block diagram of the ARP 2600 in which the different modules and their pre-wiring can be seen.

146 Synthesizers and Subtractive Synthesis 1

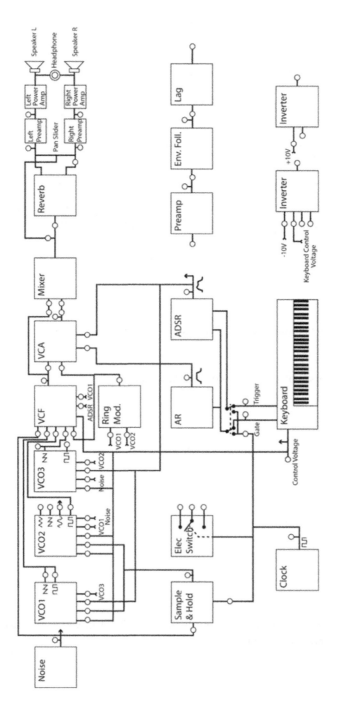

Figure 4.12. Block diagram of the ARP 2600

NOTE.– The ARP 2600 is a legendary synthesizer with only a few copies of it being produced. Getting a clone is easier today than finding a vintage machine in perfect working order. The closest replicas, such as those from Korg or synthCube, are very expensive. Turning to Behringer is less expensive, however there are even more affordable software replicas with advanced features and characteristics that more than make up for their lack of hardware. Table 4.2 presents a non-exhaustive list of some software clones of the ARP 2600[1].

Editor	Name	Compatibility	Comments
Arturia	ARP 2600 V	macOS Microsoft Windows	Paid
Sonivox	Timewarp 2600	macOS Microsoft Windows	Paid
Cherry Audio	CA2600	macOS Microsoft Windows	Paid

Table 4.2. *Virtual ARP 2600s*

4.1.2. *The Minimoog*

In 1964, the first entirely modular and relatively complex Moog synthesizers appeared. In 1969, Bill Hemsath, an engineer from Robert Moog's company, envisaged a new type of machine that was easier for a musician to implement and he built a prototype called "Min-A". However, his idea did not achieve the expected success of Bob Moog.

Figure 4.13. *The Moog Min-A, or Model A*

---

1 There are others, but unfortunately their age means that they are no longer functional with recent microcomputers.

In the following months, Hemsath continued to work on his project and, with the help of employees, built the Min-B and then the Min-C.

The appeal of large modular synthesizers was not there, and commercially it was a failure. Bob Moog had to go with the idea of Hemsath to save the company, and the prototype of the Model D, renamed Minimoog, was created in 1970. It was one of the first portable synthesizers integrating a keyboard and all the controls necessary to make sound synthesis.

In 1971, the first Model D was offered to the public. It had some modifications compared to the original: more ergonomic toggle switches, modulation wheels and pitchbends. In fact, there were three versions of the Minimoog, 1971, 1972, and 1973–1981, each with minor variations such as in the aesthetics.

Figure 4.14. *The Minimoog or Moog Model D*

The Minimoog Model D was reissued several times until the end of the 1990s when Moog Synthesizers added MIDI functions and a few other updates. In 2002, Robert Moog designed a new instrument, the Minimoog Voyager, which was available in several versions (standard, old-school, electric blue, XL, white edition, rack mount edition and select series).

NOTE.– In 2016, a few copies of the Minimoog Model D Reissue were produced by Moog. It had some improvements, including an LFO with two waveforms, triangular and square. The control knob for this LFO was located above the pitch and mod wheels. Pulling the button switched from triangular mode to square mode. On the other hand, the integrated keyboard was a high-quality Fatar TP-9 keyboard and there was a MIDI input/output, as well as an external modulation CV input and pitch and gate CV outputs.

Figure 4.15. *Standard Minimoog Voyager*

As for the ARP 2600, the Minimoog had its clones too, the Midimoog by Studio Electronics, which could be rack mounted. It contained a real Minimoog Model D card with the addition of an LFO and a MIDI port.

Figure 4.16. *The Midimoog*

Studio Electronics also created the SE-1 and SE-1X models, which were more advanced machines with additional features: four multi-stage envelope generators, ring-modulation effect, glide/auto-glide and the ability to use presets and store multiple patches.

Figure 4.17. *(a) The Minimoog SE-1 and (b) SE-1X by Studio Electronics*

Another clone was the Behringer Model D, released in 2017 as a compatible analog Eurorack[2] module. It had all the functions of an original Model D with a few additions, an LFO, a MIDI port, a USB-MIDI port, 13 patch points and the option of chaining to link 16 modules and thus have 16 voices.

Figure 4.18. *Behringer Model D*

In 2019, Behringer produced another model, the Poly D, equipped with a 37-key velocity-sensitive keyboard, four polyphonic voices, four oscillators, a 32-step sequencer, an arpeggiator, two wheels for pitchbend and modulation and a stereo chorus effect.

Figure 4.19. *Behringer Poly D*

To sum up, here are the characteristics common to all Minimoogs:

– monophonic – analog;

– three oscillators (VCO);

– one noise generator (white/pink);

– one external input;

---

2 Eurorack is a standard modular synthesizer, invented by Dieter Döpfer in 1995. It uses 3U high racks in which signals are transmitted in CV/Gate mode.

– one 24 dB/octave filter (VCF);

– two attack, decay, sustain (ADS) envelopes for VCA and VCF;

– one LFO;

– one five-voice mixer (VCO 1, 2, 3, noise, external signal);

– headphone jack;

– CV/Gate connection for oscillator, filter and tone controls.

Figure 4.20 shows an overview of the Minimoog Model D connectivity in which we can see the relationships between the different synthesizer constituents.

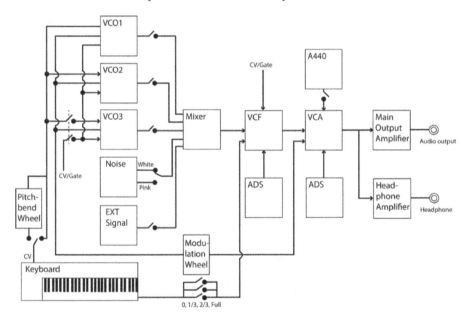

Figure 4.20. *Simplified block diagram for the Minimoog*

NOTE.– Not everyone has access to a Minimoog. An interesting alternative is to use a software version of the machine. There are several on the market. Table 4.3 shows a non-exhaustive list. You will find the links to these different products in the reference section of this book.

Editor	Name	Compatibility	Comments
Arturia	Mini V	macOS Microsoft Windows	Paid
Universal Audio	Moog LUNA	macOS	Paid
Synapse Audio Software	The Legend	macOS Microsoft Windows	Paid
Native Instruments	Monark	macOS Microsoft Windows	Paid
Gforce Software	Minimonsta	macOS Microsoft Windows	Paid
IK Multimedia	Syntronik Minimod	macOS Microsoft Windows	Paid
U-He	Diva	macOS Microsoft Windows	Paid
Moog Music Inc	Minimoog Model D Synthesizer	iOS	Paid

Table 4.3. *Software synthesizers emulating the Minimoog*

4.1.3. The Behringer Neutron

Uli Behringer founded his company in 1989. Both a musician and sound engineer, his budget did not allow him to buy the equipment he dreamed of for his own studio. So, he started by making some materials for himself and soon for his friends. From this experience, he developed the philosophy of what would become the Behringer company: to make products accessible to all, musicians, recording studios, DJs and sound and broadcasting professionals.

Since 2016, Behringer has been a manufacturer of synthesizers and drum machines, building original synthesizers but also many vintage synthesizer clones and associated electronic chips.

In 2016, Behringer released the first two synthesizers, the Deepmind 12 and the Deepmind 6. Shortly after the Deepmind 12D was released, a desktop version. A second synthesizer, the Neutron, was released in 2018 and a third, the Behringer Crave, a semi-modular synthesizer, was released in 2019. Many other instruments were released (Model D, B 2600, Monopoly, Odyssey, PRO-1, TD-3, VC 340, RD-6, RD-8, etc.).

Work Environment 153

Figure 4.21. *Some Behringer synthesizers (WASP, CAT, MS-101, VC340)*

The Neutron is not a clone but a semi-modular synthesizer, entirely designed by the manufacturer. It is paraphonic and has two oscillators (VCO), a multimode filter (VCF) (low-pass, band-pass, high-pass), two ADSR-type envelopes, an LFO, a VCA, a delay and an overdrive adjustable in saturation, tone and level.

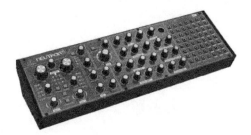

Figure 4.22. *Behringer Neutron*

The oscillators are made up of V3340 circuits, which are clones of the CEM3340, used in many vintage synthesizers like the Pro-One, Prophet 5, OB-XA, Memorymoog, SH101 and many others. Five waveforms are available: sine, triangular, sawtooth, square and pulse. Switching from one to the other can be done in discrete or continuous mode. Tuning can be done on 32, 16 or 8 feet, and fine or automatic tuning is available. The two oscillators can be synchronized, and the width of the signals can be adjusted for each.

The filter can be parameterized in frequency, modulation and resonance. Depth of modulation may vary depending on Envelope 2 and Filter.

Neutron envelopes are of the ADSR type, available for VCF and VCA in default routing.

The LFO has five wave types – sine, triangle, square, sawtooth and ramp – which can vary from 0.01 Hz to 10 kHz. This LFO can be synchronized via a MIDI clock and can modulate the cutoff frequency of the filter or pulse waves (PW).

A bias function, placed behind the VCA, provides a constant output signal. An analog delay is available at the output and one can adjust its range from 24 to 640 ms, the number of repetitions and hardness of the signal (wet/dry).

It is possible to chain several Neutron instruments to obtain polyphony.

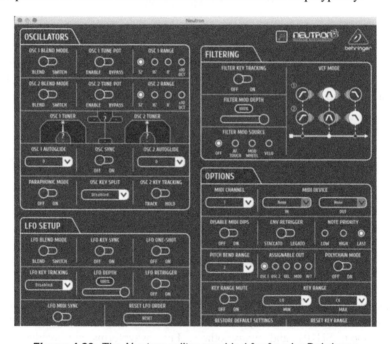

Figure 4.23. *The Neutron editor provided for free by Behringer*

An application provided by Behringer allows you to edit the functionality of the Neutron. Some parameters are only accessible via this tool.

NOTE.– There is another Neutron editor, also free, running on Ctrlr. You will find it in the reference section of this book.

To summarize, here are the characteristics of the Behringer Neutron:

– digitally controlled analog – monophonic;

– two V3340 circuit oscillators (VCO) – five waveforms;

– paraphonic mode (two voices);

– multimode filter (VCF) with resonance;

– two analog ADSR envelopes for VCF and VCA;

– one LFO with five waveforms, synchronizable via MIDI;

– one noise generator;

– bucket-brigade device (BBD) analog delay;

– overdrive circuit;

– 32 input patch points;

– 24 output patch points;

– MIDI IN/THRU;

– USB MIDI;

– 6.35-mm jack external sound input;

– 6.35-mm jack headphone output;

– 6.35-mm jack line output;

– Eurorack format;

– dimensions: 94 × 424 × 136 mm;

– external power supply (12 V CC, 1,000 mA);

– weight: 2 kg.

Figure 4.24 shows the block diagram of the Neutron in which different modules and their pre-wiring can be seen.

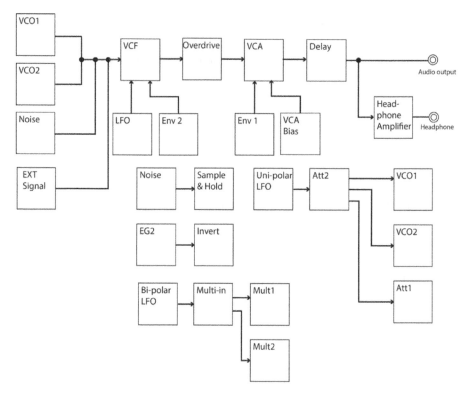

Figure 4.24. *Default Neutron pre-wiring (routing)*

4.1.4. *The Novation Bass Station II*

Novation Digital Music Systems is a British manufacturer of musical equipment. The company was founded in 1992 by Ian Jannaway and Mark Thompson and manufactured synthesizers, several types of MIDI controllers, keyboards or pads and audio interfaces.

The first product released by Novation in 1992 was the MM10, a MIDI keyboard controller designed to work with the Yamaha QY10 workstation. The concept was innovative for the time; it was one of the first, if not the first, fully portable 8-track digital audio workstations (DAWs).

In 1993, the company launched the Novation Bass Station, a compact synthesizer intended for instrumental accompaniment built around two digitally controlled analog oscillators (DCOs), an LFO and a filter.

In 1995, Novation introduced Analogue Sound Modelling (ASM) technology with the Drum Station that emulated the famous Roland TR-808 and TR-909 drum machines.

It was Chris Huggett, designer of the OSCar, WASP synthesizers and developer of the sampler operating system Akai S1000, who advised on and participated in the creation of the Bass Station and the Drum Station.

Huggett eventually joined Novation Digital Music in the mid-1990s and designed the Novation Supernova, a multitimbral polyphonic synthesizer in 1998.

The company later created audio interfaces, then controllers, eventually turning to software applications.

In 2004, Novation Digital Music Systems was acquired by Focusrite.

Many other products emerged, including Launchpad, Ultranova, MiniNova, Peak and Launchpad X and pro.

It was in 2013 that the Novation Bass Station II was created, designed around three oscillators (two DCOs + one Sub DCO), two filters, two LFOs, an effects section, a patch backup, a sequencer, an arpeggiator, a keyboard with 25 keys sensitive to velocity and pressure and a modulation section. The set is completed by a MIDI interface, USB connectivity and an AFX mode allowing for the modification of the patches.

This small synthesizer has enormous potential in terms of sound editing and a very loud sound. Its compactness is detrimental to the ergonomics in terms of use as many functions share the same commands, and some are only accessible by pressing the Function key and a key on the keyboard.

The oscillators are very stable, with a range of 2–16 feet, tuning on two octaves to a tenth of a tone, fine tuning to a hundredth of a tone, and possible synchronization, and the generated signal is entirely analog, although it is digitally controlled. It should be noted that the value of the chord can fluctuate, thus allowing for the simulation of a vintage VCO. The Sub DCO, which covers two octaves, is controlled by DCO1. The two DCOs can be ring modulated.

All of the signals, DCO1, DCO2, Sub, Ring Modulator, noise and external signal, are mixed and then directed to a multimode filter. The Classic mode works in low-pass band-pass or high-pass on two or four poles[3], which provides significant resonance and aggressive filtering that can go as far as self-oscillation. Acid mode activates a four-pole low-pass that gives the typical 1980s sound with a clean resonance and restrained frequency response.

Overdrive before filtering and distortion after filtering are available to add additional consonance and resonances.

Behind the filter section, the signal is directed to a VCA equipped with a limiter, before reaching a monophonic 6.35-mm jack output.

Figure 4.25. *Novation Bass Station II*

On the modulation side, two LFOs with four waveforms, a variable speed and a delay are available. Two ADSR-type envelopes can be triggered in three modes: single, multiple or automatic. Attack ranges from 0 to 5 s and release extends up to 10 s. The portamento also benefits from three modes.

It should be noted that the modulation wheel can be assigned to the filter cutoff and the pitch of the DCO2. Pressing the keyboard can control filter cutoff and LFO2 speed. As for the velocity, it can interact on the envelope, the pitch, the volume and the filter.

The Bass Station II includes a sequencer and an arpeggiator that can be synchronized via an internal clock or MIDI. There are 32 arpeggio patterns that take up sophisticated rhythms. The speed is adjustable to set the desired tempo. It works on one to four octaves with six different modes (high, low, random, etc.).

---

3 The number of poles indicates the attenuation of the filter: one pole = 6 dB/octave, two poles = 12 dB/octave, three poles = 18 dB/octave and four poles = 24 dB/octave.

The sequencer is programmable step by step up to 32 notes and can record up to four sequences stored in memory that are always accessible, even after stopping the synthesizer. Sequences are triggered from the keyboard and can be transposed. In agreement with the arpeggiator, the 32 variations of the latter are accessible to modify the temporality of the sequence.

In order to edit the Bass Station II parameters more easily, there are some software editors on the market. Table 4.4 summarizes some of them.

Editor	Name	Compatibility	Comments
Patch Base CoffeeShopped LLC	Novation Bass Station II Editor and Librarian	iPad and macOS	Paid
StudioCode.dev	Bass Station 2 Editor	Online	Free
Synthmania	Novation Bass Station 2 Editor	macOS Microsoft Windows	Paid
Filthy Filterz	BS2 Editor	macOS Microsoft Windows	Paid
Joe Mattiello (github.com)	Bass Station II Max For Live	macOS	Free, works with Ableton Live
Midiquest	Novation Bass Station II Editor Librarian	macOS Microsoft Windows	Paid

Table 4.4. *Editors for the Novation Bass Station II*

To summarize, here are the features of the Novation Bass Station II:

– digitally controlled analog;

– 25-note velocity-sensitive keyboard, aftertouch assignable to filter frequency, LFO or VC2;

– three oscillators (two DCO and a Sub) with four waveforms (sinusoidal, triangular, sawtooth, variable square);

– two filters (ladder and classic derived from the original Bass Station);

– two ADSR-type envelopes;

– two synchronizable LFOs with four choices (triangular, sawtooth, square, sample and hold);

– one noise generator;

– one arpeggiator/sequencer with 32 modes over four octaves (direction and swing function);

– portamento;

– modulation wheel assignable to LFO1, LFO2 or VCO2;

– pitchbend wheel;

– 64 factory presets;

– 64 user presets;

– USB2 MIDI;

– MIDI in/out;

– 6.35-mm jack mono output;

– 6.35-mm jack headphone output;

– 6.35-mm jack external input;

– external power supply or via USB port;

– dimensions: 457 mm (18″) width × 273 mm (10.75″) depth × 76 mm (3″) height.

Figure 4.26 shows the block diagram of Bass Station II, with the dotted arrows indicating the possible modulation paths. Sub DCO is linked to DCO1.

4.1.5. *The Arturia MatrixBrute*

Arturia is a French company founded in 1999 by Frédéric Brun and Gilles Pommereuil. The company's goal was to produce software synthesizers. The first product developed was a virtual instrument workstation.

In 2003, Arturia worked with Robert Moog to create the Modular V virtual synthesizer, which emulated the Moog 3C and Moog 55 modules. Building on its success, other models were quickly marketed, the ARP 2600, the Minimoog, the Roland Jupiter-8 and the Sequential Circuit Prophet 5.

In 2008, a hybrid solution combining a MIDI keyboard controller and a set of synthesizer patches was presented by the company under the name Analog Experience.

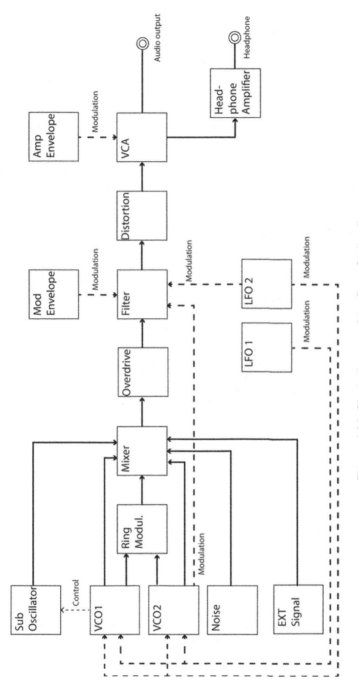

Figure 4.26. *Block diagram of the Bass Station II*

The year 2009 saw the advent of Origin Desktop, a polyphonic system incorporating previously developed software synthesizers designed around DSP. It provided software emulations of classic synthesis modules (VCO, VCF, VCA, Mixer, Ring Modulator, etc.) and virtual models (ARP 2600, Yamaha CS-80, Roland Jupiter 8, Sequential Circuit Prophet 5, Minimoog, etc.). The first version was a tabletop instrument. The second version included the Origin Keyboard.

It was not until 2012 that Arturia began to manufacture hardware synthesizers, their first model being the MiniBrute, a monophonic analog synthesizer with a 25-key keyboard sensitive to velocity and pressure, integrating a VCO, two LFOs, a Steiner-Parker type filter and an arpeggiator. The MiniBrute had a CV/Gate I/O and MIDI connection. Conversion of MIDI signals to CV/Gate is available.

The following year, the MicroBrute made its appearance, a more watered down version of the MiniBrute. In 2018, it evolved into the MiniBrute 2 and then the MiniBrute 2S, which abandoned its keyboard in favor of pads.

In 2016, the MatrixBrute made its debut, followed by the Microfreak in 2019, a hybrid memory synthesizer and then the PolyBrute, a high-end polyphonic synthesizer.

Figure 4.27. *Arturia PolyBrute*

In parallel, Arturia continued to develop many other software and hardware products, software emulations of various synthesizers, iOS synthesizers, MIDI controllers, software effects, audio interfaces, drum machines and so on.

Returning to the MatrixBrute, this machine is still, six years after its release, one of the most sophisticated on the market, a real beast of a machine with extraordinary features.

Let us start with the three oscillators, which can be tuned in semitones over two octaves. A fine-tuning function is present to fine-tune the settings. Each VCO1 and VCO2 has three main waveforms – sawtooth, square and triangular – which can be more complex in ultrasaw, pulse width and metalizer mode to give them more richness. A sub-oscillator mode, one octave lower, is also present. The third oscillator has four waveforms and can be transformed into an LFO, which divides the original frequency by four in order to become a modulator.

Figure 4.28. *Arturia MiniBrute*

Its strong point is how the oscillators interact with each other. VCO1 can be routed to VCO2 to modulate it (frequency modulation, FM), while VCO3 can modulate both VCO1 and VCO2. Filters can also be modulated, VCO3 on VCF1 and VCF2. VCO2 can force VCO1 to follow its pitch (VCO Sync). The noise generator can also modulate VCO1 or VCF1.

Figure 4.29. *Arturia MatrixBrute*

Let us examine the latter. The noise can have four different colors: white, pink, red and blue. Blue noise is more energetic in the high frequencies, and it is followed by the more serious white noise, and then pink noise and red noise, which favor the low frequencies even more.

As for the filters, the MatrixBrute has two independent ones. Each of the sources, VCO, noise generator and external input, can be assigned to one of the filters or to both at the same time, either in series or in parallel. The filters are different: the first is a multimode Steiner-Parker filter (low-pass, high-pass, band-pass and rejector) with two or four poles; the second is a ladder filter, the famous filter invented by Robert Moog, also multimode (low-pass, high-pass, band-pass) with two or four poles. Filter cutoff frequencies can be controlled simultaneously.

For modulations, there are multiple sources; they can act on the VCFs or the VCA, but other assignments are possible. The three LFOs, the third being VCO3, have several waveforms and can be set to a chosen frequency between 0.06 and 100 Hz, or synchronized to the tempo. Their triggering can be free or on each note. The LFO1 phase is assignable as well as the delay for LFO2.

Behind the filters, the signal is directed to a VCA itself, followed by a set of analog effects, stereo delay, mono delay, chorus, flanger and reverb. Each effect can be set independently.

The three ADSR envelopes have times that can vary between 2 ms and more than 10 s, which offers many possibilities. An envelope is assigned to the cutoff frequency of the filters according to an adjustable quantity for each. The level can be modulated via keyboard velocity. A second envelope is routed by default to the VCA, whose level can also be modulated via velocity. The third envelope is assignable and has an adjustable delay, with its assignment being defined by the matrix.

This 16 × 16 dot matrix is one of the main elements of the MatrixBrute. The sources are in rows (from A to P) and the destinations in columns (from 1 to 16). The connection, the equivalent of a patch cable on a modular or semi-modular synthesizer, is done simply by pressing the button located at the intersection, which instantly lights up pink.

The amount of modulation is determined by the encoder above it and can vary from −99 to +99. Negative values decrease the amount of modulation of a destination as the input voltage increases. Positive values, on the other hand, increase the modulation. When you go back to pre-selection mode, the buttons turn blue. Four encoders, called Macro encoders, can act at the same time on the matrix, thus allowing for the modification of several parameters.

Let us continue with the sequencer and the arpeggiator. The sequencer offers 64 configurable steps according to four variables – activation, slide, modulation and accent. The recording of the pattern can be done step by step or in real time. Reading is done in several directions (forward, backward, alternate and random) according to four temporal divisions (1/4, 1/8, 1/16 and 1/32) and three note durations (normal, triplet and dotted). The tempo is variable or synchronized to the sequence via MIDI. Using the keyboard, a sequence can be transposed in real time. The arpeggiator cycles through the held notes one after the other up to a maximum of 16. It responds to velocity. The sequence of notes can be set according to five combinations (up, down, up/down, random and octave). There is a special mode, called the Matrix Arpeggiator, which allows up to four notes to be held down and played in any order, following a defined octave for each note over 16 steps. It is also possible to vary the rhythm.

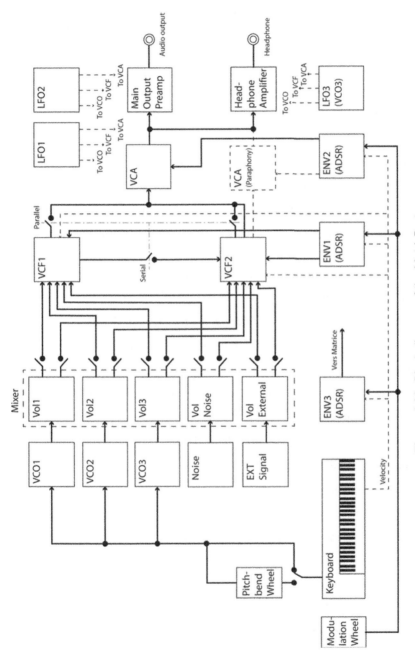

Figure 4.30. Block diagram of the MatrixBrute

Finally, note that the MatrixBrute, in addition to the presence of standard MIDI and MIDI-USB connectors, which can receive or transmit MIDI CCs, has 12 CV/Gate inputs/outputs to communicate with other machines such as modular synthesizers.

Figure 4.30 shows a simplified block diagram of the MatrixBrute on which the matrix does not appear but instead the default pre-wiring of the different sections.

To summarize, here are the general technical characteristics:

– digitally controlled analog;

– four-octave keyboard sensitive to velocity and pressure;

– 256 memory locations for presets;

– two oscillators (VCO) with three waveforms (sawtooth, square and triangular);

– one oscillator (VCO/LFO) with four waveforms (sawtooth, square, triangular and sine);

– one noise generator (white, pink, red and blue);

– one mixer with five inputs;

– one Steiner-Parker 12 or 24 dB/octave multimode filter (VCF);

– one ladder 12 or 24 dB/octave multimode filter (VCF);

– two VCA;

– two LFO (sine, triangular, square, ramp, sawtooth, random, sample and hold);

– two ADSR envelope generators;

– one delay, attack, decay, sustain, release (DADSR) type envelope generator;

– one external audio input;

– one 16 × 16 modulation matrix;

– one 64-step sequencer;

– one effect generator (delay stereo, delay, flanger, chorus and reverb);

– one arpeggiator;

– 12 CV/Gate input/output;

– one 6.35-mm jack stereo output;

– one MIDI input/output (IN, OUT, THRU);

– one USB-B MIDI input/output;

- two pedal inputs (expression and sustain);
- dimensions : 855 × 112 × 365 mm;
- weight 7.9 kg.

4.2. Software

After reviewing the different types of hardware were selected, here is the chosen sound synthesis software. Native Instruments Reaktor and Cycling '74 Max/MSP are paid, whereas Pure Data and VCV Rack are free[4].

The hardware presented above offers very advanced possibilities, and the software does not disappoint. The multiple and advanced functionalities make it possible to create synthesizers working in subtractive synthesis, although it is possible to create machines in all the other modes of synthesis, or even hybrid machines.

As for the hardware, the selected software has an unfailing reputation among professionals and enlightened sound synthesis amateurs.

Native Instruments Reaktor and VCV Rack are very similar to modular hardware synthesizers, faithfully retaining their configuration with a set of modules connected by connecting cables, often called patch cables. Max/MSP and Pure Data are above all visual sound programming tools which, in addition to their ability to develop synthesizers, also offer much wider possibilities.

4.2.1. Native Instruments Reaktor

In 1996, Native Instruments launched Generator, the predecessor of Reaktor. It was a software synthesizer that ran on Microsoft Windows on PC. It used an interface that allowed users to combine various blocks (oscillators, envelopes, filters, LFOs, mixers, MIDI sources, delays, etc.) to create your own synthesizer.

For beginners, it also offered a library of ready-to-use instruments (analog synthesizers, FM synthesizers, organs, strings, percussion, sound effects, etc.). There were numerous possibilities and the software was rather revolutionary for its time, although it required a powerful computer to fully exploit all its functions.

---

4 The user needs to pay for VCV Rack 2.0 to use certain functionalities.

Three years later, in 1999, Native presented Reaktor 2, with PC/Macintosh compatibility. Several plugin formats were included. Very quickly, many users developed synthesizers, drum machines, sound effects and a whole set of exotic instruments, favoring the creation of a huge library of musical products.

Reaktor version 3 was released in 2001 with new modules, an improved audio engine and full cross-platform compatibility.

For version 4, which arrived on the market in 2003, the sound quality was further improved, graphic signal controllers were present, the library of instruments had broadened even more and the interface had been revised.

Version 5, in 2005, brought "Core Technology", a powerful programming system and "Core Cell" modules. At this stage, the software was powerful and offered many possibilities. It had the power and sound quality but mastering it required a significant investment in terms of time. It should be noted that version 5.5, with some notable improvements, was released in 2010.

In 2015, Reaktor 6 brough improved user interface ergonomics and new modules including recent synthesizers and the notion of blocks that simplified the modular approach.

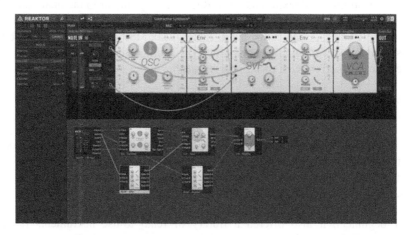

Figure 4.31. *A simple synthesizer built with Reaktor Blocks*

4.2.2. VCV Rack 2

The VCV Rack is a free or paid (advanced options) cross-platform (Linux, macOS, Microsoft Windows) virtual analog modular synthesizer that simulates a

Eurorack where, as in reality, users can connect different modules to each other by cables.

In 2016, Andrew Belt founded VCV in the United States, and in 2017, during Knobcon, a convention dedicated to synthesizers, the VCV Rack was presented.

Independent developers can design free or paid modules; to date, there are several thousand of them, including a few dozen from official Eurorack module manufacturers.

The software is compatible with Eurorack hardware modules connected via a USB or MIDI interface. Synchronization between hardware and software is possible.

The VCV Rack can also be integrated as a VST plugin and vice versa, VST plugins can be added to a VCV rack via a VCV Host.

In June 2019, polyphony was considered, and multiprocessor management became possible.

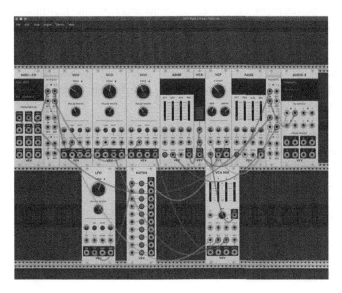

Figure 4.32. *An example of a virtual Eurorack on a VCV Rack 2*

Today, the VCV Rack has evolved into the VCV Rack 2.

The VCV has an active community of developers scattered around the world.

A paid version of the VCV Rack 2 is available and brings additional features.

4.2.3. *Cycling '74 Max/MSP*

Max is a visual programming language developed by Miller Puckette at IRCAM in the mid-1980s. Initially, it was called Patcher. It was the Opcode System company that marketed it from the 1990s, with Cycling '74 having taken over since 1999.

A very large community of developers and users have used it successfully. Also well used by composers, musicians and electronics engineers, Max/MSP is a popular tool for musical research, live performances, the design of digital works of art and for performances of all kinds.

Over the years, Max/MSP has evolved, its graphical interface has become more ergonomic, debugging aids have been integrated, multiprocessing management has been added and so on.

Max/MSP was born from the combination of two pieces of software. Max was geared more towards mathematical calculations and real-time instrument control via MIDI, and MSP, a library of particularly powerful functions, focused on audio signal processing.

Its operation is based on the creation of a "patch" by the user, hence its historical name, which makes it possible to design and formalize an audio or musical processing project.

In a dedicated work window known as the patcher – a blank page – the user can drag objects of different types (messages, mathematical operators, controllers, filters, buttons, etc.) with a defined function. A patch, the content of the patcher, is made up of objects whose inlets and outlets make it possible to generate, receive or disseminate information and values, thus constituting processing tools of different kinds (MIDI signals, audio, etc.).

The patch can be exported in a so-called "stand-alone" mode in order to offer future users an exploitable version of the project without them needing to have a commercial license for Max/MSP.

A set of additional modules such as Jitter and Vizzie can be used with Max/MSP, which open the way to image processing, matrix analysis, sound effects or real-time video.

Max/MSP runs on macOS and Microsoft Windows environments. In partnership with Ableton, Cycling '74 developed a specific version of Max/MSP, named Max for Live, for the Ableton Live software music sequencer, a live-oriented composition and arrangement tool. Max for Live provides access to the Max development environment directly within Live. It offers a whole range of musical instruments and effects. The Live user can also create or customize their peripherals.

Finally, we can ask ourselves why the name "Max/MSP" was chosen. It was chosen as a tribute to Max Mathews, one of the pioneers and fathers of computer music. As for MSP, there are two possible explanations, the acronym of Max Signal Processing or that of Miller Smith Puckette.

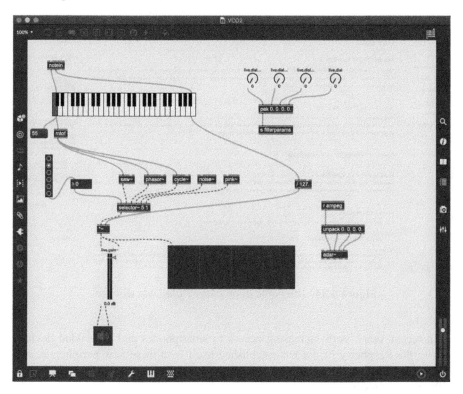

Figure 4.33. *Max/MSP patcher window and a patch connecting multiple objects*

4.2.4. Pure Data

In 1996, Miller Puckette, the developer of Max/MSP, continued to work on his project and created Pure Data, also called "Pd", a free version[5] of its software.

---

5 Free download available at: www.puredata.info/.

Distributed freely on the Web[6], it is available for Linux, macOS, and Microsoft Windows platforms. As for Max, a Pd community exists; it is very active and responsible for constant evolutions of the software and many projects.

Pd has many similarities with Max/MSP although its evolution and ergonomics are different; it is a visual programming language based on identical foundations.

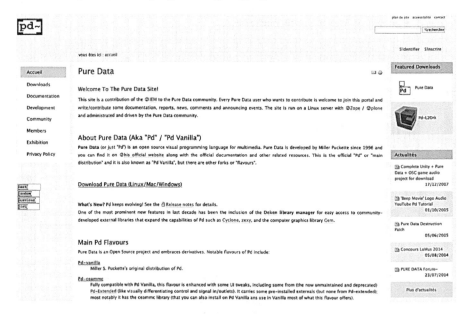

Figure 4.34. *The home page of the Pure Data website*

In recent years, with increasing access to smartphones and embedded devices such as the Raspberry Pi, Pd has established itself with these new peripherals, and specific libraries have emerged such as libpd, DroidParty for Google Android and PdParty for Apple iOS.

The processing of sound, video and 2D/3D graphics, and the interfacing of sensors, input-output peripherals and MIDI devices are within the reach of Pd. It can work on networks, control motorized systems, lighting systems and many other types of multimedia equipment.

In Pd, algorithmic functions are represented by visual boxes, called objects, placed in a patch window called a canvas. Data flow between objects is achieved via visual connections called patch cords.

---

6 Available at: www.puredata.info.

Each object performs a specific task, which can vary in complexity, from low-level mathematical operations to complex audio or video functions such as reverberation, FFT transformations, or video decoding.

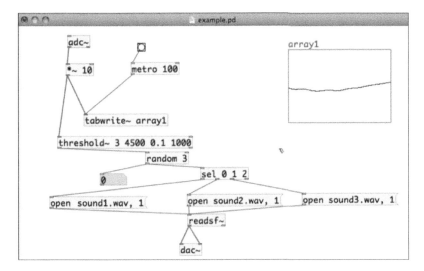

Figure 4.35. *A Pd canvas containing several objects*

4.3. Conclusion

In this chapter, you discovered five hardware synthesizers, ARP 2600, Minimoog, Behringer Neutron, Novation Bass Station II and Arturia MatrixBrute, and four sound synthesis tools. As I have already specified, these were presented so the reader can have a precise idea of subtractive synthesis within sound systems with different architectures.

In Chapters 1–5 of Volume 2, you will be able to gradually discover, by means of exercises, subtractive synthesis in its simplest forms up to the most advanced uses.

5
CV/Gate and MIDI

The CV/Gate mode and the Musical Instrument Digital Interface (MIDI) standard are attached to synthesizers, without which they could not communicate with other peripherals such as keyboards, sequencers and various controllers.

Although the CV/Gate mode dates back to the late 1970s, it is still widely used today. As for MIDI, which appeared in the mid-1980s, no modern synthesizer seems to be able to do without it, with this protocol being one of the keys to communication between the various electronic machines in the musical universe.

While CV/Gate and MIDI seem very different, they are not completely disconnected; devices for switching from one to the other are presented in section 5.3.

5.1. CV/Gate

CV/Gate is an analog control mode for synthesizers and some other equipment like sequencers or drum machines.

Via a control voltage (CV), the pitch of a note is modified and the control signal of a gate activates or deactivates it.

5.1.1. *Overview*

This technique was developed in 1977 and continues to be used today. However, it was replaced in 1983 by the MIDI standard, which is more precise and has much greater functionality.

Many devices still use CV/Gate mode alongside MIDI to run old and new equipment.

This mode was particularly suitable for the analog and monophonic synthesizers of the 1970s–1980s, which had voltage-controlled oscillators and voltage-controlled filters.

5.1.2. *Operation*

The voltage control specifies the note to be played with a different voltage for each. It can also control additional parameters such as a duration, for example.

A gate, also called a trigger, produces the note; it can also trigger an event, often associated with an ADSR-type envelope.

Unfortunately, not all manufacturers use the same voltage splitting to control their instruments and several CV/Gate modes exist:

– For Moog, ARP, Roland and Oberheim and Sequential Circuits, an octave spans a voltage range equal to 1 V. This standard, "volts per octave", was defined by Robert Moog in the 1960s. The voltage will be between 0 and a maximum of +5 volts.

– For other manufacturers, DC voltages of –5 to +5 volts or 0 to 10 volts have been used.

– Korg and Yamaha use "hertz per volt", doubling the voltage represents one octave in pitch.

Several remarks must be made regarding the following elements:

– the wiring can be different with a positive voltage on the tip or on the ring of a 6.35-mm stereo jack socket, also called tip ring sleeve (TRS);

– a 6.35-mm mono tip sleeve (TS) jack connection can also be used;

– some equipment from Korg and Yamaha does not work in hertz/volts;

– there are other possible variations of an octave, for example 1.2 volts/octave in order to have 0.1 volts for a semitone.

5.1.3. *Note definition*

Let us determine the voltage of each note for the two types of CV/Gate, "volts per octave" and "hertz per volt".

Voltage (V)	Semitone	Note	Octave	Frequency (Hz)
0.250	1	C	0	32.703
0.333	2	C#	0	34.648
0.417	3	D	0	36.708
0.500	4	D#	0	38.891
0.583	5	E	0	41.203
0.667	6	F	0	43.654
0.750	7	F#	0	46.249
0.833	8	G	0	48.999
0.917	9	G#	0	51.913
1.000	**10**	**A**	**0**	**55.000**
1.083	11	A#	0	58.270
1.167	12	B	0	61.735
1.250	1	C	1	65.406
1.333	2	C#	1	69.296
1.417	3	D	1	73.416
1.500	4	D#	1	77.782
1.583	5	E	1	82.407
1.667	6	F	1	87.307
1.750	7	F#	1	92.499
1.833	8	G	1	97.999
1.917	9	G#	1	103.826
2.000	10	A	1	110.000
2.083	11	A#	1	116.541
2.167	12	B	1	123.471
2.250	1	C	2	130.813
2.333	2	C#	2	138.591
2.417	3	D	2	146.832
2.500	4	D#	2	155.563
2.583	5	E	2	164.814
2.667	6	F	2	174.614
2.750	7	F#	2	184.997
2.833	8	G	2	195.998
2.917	9	G#	2	207.652
3.000	10	A	2	220.000
3.083	11	A#	2	233.082
3.167	12	B	2	246.942
3.250	1	C	3	261.626
3.333	2	C#	3	277.183
3.417	3	D	3	293.665
3.500	4	D#	3	311.127
3.583	5	E	3	329.628
3.667	6	F	3	349.228
3.750	7	F#	3	369.994
3.833	8	G	3	391.995
3.917	9	G#	3	415.305
4.000	10	A	3	440.000
4.083	11	A#	3	466.164
4.167	12	B	3	493.883

Table 5.1. *Notes in volts per octave for octaves 0 to 3*

5.1.3.1. *The volts per octave standard*

For a voltage of 1 volt/octave, Table 5.1 gives an example of note distribution starting from 55 Hz for 1 V.

NOTE.– The frequency of a note can be calculated with the following formula:

$$f_{note} = f_{ref} \times 2^{(octave-3)+\frac{semitone-10}{12}},$$

where:

– f_{ref} is the reference frequency, or 440 Hz, corresponding to La (A);

– an octave is an integer between 1 and 9. The A (A) 440 Hz is in octave 3 or 4 depending on conventions – for octave 4, octave –3 must be changed to octave –4 in the formula;

– a semitone is an integer between 1 and 12. The A (A) 440 Hz is tone 10 on the scale.

Let us take an example: for an Re#1 (D#1), with an La (A) in octave 3, we have:

$$440 \times 2^{(1-3)+\frac{4-10}{12}} = 77.782 \text{ Hz}.$$

CALCULATION.– An octave is equal to 12 semitones and 1 V, so 1/12 V is equal to 1 semitone; it is therefore sufficient to multiply 1/12 V, or approximately 0.0833 V, by the number of semitones desired and add or subtract from it the voltage of the starting note to obtain the voltage for the desired note.

For example: if La1 (A1) = 2 V, what is the voltage corresponding to Mi2 (E2)? La1 to Mi2 (E2) => 7 semitones, the voltage should be 7 × 1/12 + 2 = 2.583 V.

5.1.3.2. *The hertz per volt standard*

For a frequency of 55 Hz/V, Table 5.2 gives an example of note distribution if 55 Hz is obtained with a control voltage of 1 V.

CALCULATION.– To calculate the voltage as a function of the frequency, simply divide the latter by 55.

For example: if the note is Re2# (D2#), which has a frequency of 155.563 Hz, the voltage should be 155.563/55 = 2.828 V.

Voltage (V)	Semitone	Note	Octave	Frequency (Hz)
0.595	1	C	0	32.703
0.630	2	C#	0	34.648
0.667	3	D	0	36.708
0.707	4	D#	0	38.891
0.749	5	E	0	41.203
0.794	6	F	0	43.654
0.841	7	F#	0	46.249
0.891	8	G	0	48.999
0.944	9	G#	0	51.913
1.000	**10**	**A**	**0**	**55.000**
1.059	11	A#	0	58.270
1.122	12	B	0	61.735
1.189	1	C	1	65.406
1.260	2	C#	1	69.296
1.335	3	D	1	73.416
1.414	4	D#	1	77.782
1.498	5	E	1	82.407
1.587	6	F	1	87.307
1.682	7	F#	1	92.499
1.782	8	G	1	97.999
1.888	9	G#	1	103.826
2.000	10	A	1	110.000
2.119	11	A#	1	116.541
2.245	12	B	1	123.471
2.378	1	C	2	130.813
2.520	2	C#	2	138.591
2.670	3	D	2	146.832
2.828	4	D#	2	155.563
2.997	5	E	2	164.814
3.175	6	F	2	174.614
3.364	7	F#	2	184.997
3.564	8	G	2	195.998
3.775	9	G#	2	207.652
4.000	10	A	2	220.000
4.238	11	A#	2	233.082
4.490	12	B	2	246.942
4.757	1	C	3	261.626
5.040	2	C#	3	277.183
5.339	3	D	3	293.665
5.657	4	D#	3	311.127
5.993	5	E	3	329.628
6.350	6	F	3	349.228
6.727	7	F#	3	369.994
7.127	8	G	3	391.995
7.551	9	G#	3	415.305
8.000	10	A	3	440.000
8.476	11	A#	3	466.164
8.980	12	B	3	493.883

Table 5.2. *Notes in hertz per volt*

5.1.4. *Operation of the gate (or trigger)*

It produces an all or nothing logic signal. Active values can be of different types:

– S-Trigger (short circuit trigger): closing a contact. At rest, the control input voltage is positive. This voltage becomes zero when activated.

– V-Trigger (voltage trigger): the control is carried out because of the presence of a positive voltage, which can vary between +2 and +15 V (usually +5 V) compared to ground. There is therefore an input voltage of practically zero at rest that becomes positive during activation.

5.2. Musical Instrument Digital Interface

The MIDI standard has revolutionized the field of musical equipment. It led to radical change that pushed musicians toward a new way of working, thereby influencing their creative process.

Section 5.2 discusses this standard, its evolution and its principle from its creation until today, as well as alternative and proprietary formats, GS (Roland) and XG (Yamaha).

5.2.1. *MIDI version 1.0*

One of the founding fathers of MIDI is undoubtedly Dave Smith, the founder of Sequential Circuits. Having electronic musical equipment communicate digitally, such as a synthesizer and a sequencer, was not entirely his idea. Many others had already thought of it, but everyone developed their own communication device without ever consulting or collaborating with one another.

Dave Smith had something much more ambitious in mind: to create a digital interface based on a universal communication protocol that could be used by all musical equipment regardless of manufacturer or specificities.

At the 1982 National Association of Music Merchants (NAMM) show in Anaheim, leading hardware manufacturers including Sequential Circuits, Yamaha, Roland, Korg, E-mu, Oberheim and Kawai came together to develop a protocol and communication technology using adequate, reliable and secure transfer speeds. The use of optocouplers was one of the important points raised in the debate. It allowed for the separation of the different hardware from one other and solved several problems.

Another key feature was the cost of the device. The interface had to remain inexpensive if it was to be accessible to all manufacturers.

Based on these requirements, Dave Smith and Chet Wood proposed a digital interface, which they named Musical Instrument Digital Interface. In December 1982, the first MIDI synthesizer, the Prophet 600 by Sequential Circuits, appeared on the market, closely followed by the DX7 by Yamaha, Jupiter 6 by Roland, and others.

At the NAMm show the following year, in 1983, demonstrations of MIDI communication between equipment of different brands were presented.

These first experiments, although functional, posed several technical problems that had to be solved for MIDI to become stable. In 1985, the standard was considered to have become reliable, and it continued to evolve.

For MIDI to progress, an association – the International MIDI Association – was created in 1983. MIDI 1.0 specifications were published through this organization.

midi

Figure 5.1. *The MIDI standard logo*

Alongside this organization, two other associations were created in 1985 to oversee the development of the standard and to avoid anarchy: the MIDI Manufacturers Association (MMA), which brings together American, Canadian and European manufacturers, and the Japanese MIDI Standard Committee (JMSC), which brings together manufacturers in Japan.

5.2.2. *MIDI Version 2.0*

In 2019, MMA announced MIDI version 2.0, which was officially launched in January 2020 at the NAMM show.

(X)MIDI

Figure 5.2. *The MIDI version 2.0 logo*

Several new features arrived:

– The MIDI-CI (capability inquiry), an option that extends MIDI functionality while maintaining backward compatibility with MIDI 1.0 and therefore older MIDI devices. This device makes it possible to distinguish recent products from old products. To know what data are supported, it sends a test during activation in order to best configure the MIDI environment in relation to the devices present.

– A two-way connection is possible between the devices on a single cable, whereas it required two in MIDI 1.0. MIDI 2.0 devices can communicate via the MIDI-CI to automatically configure themselves. They can also share their configurations and supported MIDI functions. If a device does not support the new features, it will usually work in MIDI 1.0.

– An accuracy that goes from 7 to 32 bits, offering a controllable velocity on 16,384 levels instead of the 128 on MIDI 1.0, and the dynamics are improved. The 128 levels for functions such as volume are now increased to several billion, thus offering a feeling and a smoothness close to that of analog. More than 32,000 controllers are available, including individual controls for each note, which makes it easier to use.

– Profile management to configure new hardware is available. It retrieves the preset names and those of each individual parameter.

– Mode management, very important in MIDI, is integrated into a list of modes and resources called ModeList. CurrentMode retrieves or sets the current mode.

– A list of programs (ProgramList) provides the list of available programs among a collection of programs. A collection groups programs with common characteristics such as type of instrument, presets and type of synthesis.

– The hardware state is known instantly via Property Exchange Resources to know the current content of the memory of a MIDI device (current program, mode, effects, etc.).

– The number of channels, limited to 16 in MIDI 1.0, increases to 256 through 16 groups of 16 channels, maintaining compatibility with the old standard by working on a single channel.

– The latency is reduced and the delay between the note played and the note heard is reduced.

– Backwards compatibility with MIDI 1.0 ensures the operation of all MIDI 1.0 hardware already on the market.

– A new universal data format, Universal MIDI Packet, makes it possible to easily transport MIDI 2.0 standard data via media such as USB or Ethernet.

NOTE.– Regarding the connectors and the type of cable, MIDI 2.0 is a transport protocol; the choice of the connection device is therefore not defined, the data may, as is already the case, pass through many connections (DIN, DB25, Firewire, TRS, Ethernet, USB, etc.).

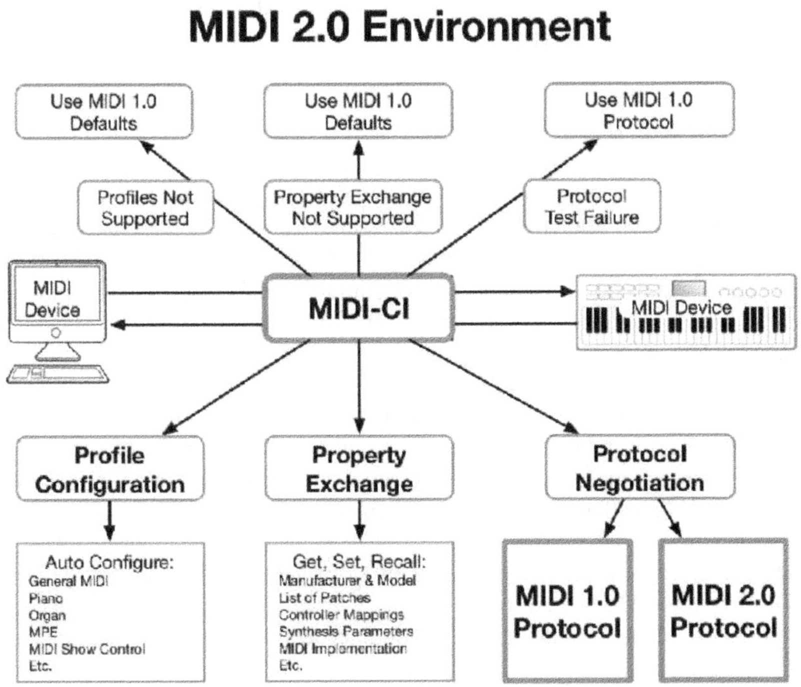

CV/Gate and MIDI 183

MIDI 2.0 Environment

Figure 5.3. *The MIDI 2.0 environment (source: www.midi.org)*

The previous lines give a brief overview of the new features and possibilities from MIDI 2.0, and manufacturers must familiarize themselves with this new standard, which should gradually be implemented on new equipment.

MMA and AMEI (Association of Musical Electronics Industry) collaboration will continue to ensure compatibility and future development of MIDI specifications.

5.2.3. *Principle*[1]

When we talk about MIDI, we must make the distinction between two important elements, the hardware part, made up of electronic components that allow for the connection of musical devices to one another, and the software part based on a

---

[1] We will detail here the operation of version 1.0 of the MIDI standard, which is still widely used.

communication protocol that ensures the coding, decoding and data transport as a MIDI message based on the MIDI language.

5.2.4. *The hardware*

The MIDI interface, like any means of digital data transmission, is characterized by its speed or bitrate, that is 31,250 bits/s. Each datum transmitted has a length of 8 bits, an octet, with a start bit and a stop bit without a parity bit (UART[2] over 10 bits). The transmission is asynchronous to limit the number of transmission wires within the transport cable and to avoid the time lag that can appear with a synchronous device.

A MIDI link ensures the transmission of information in a single direction from the transmitting equipment to the receiving equipment. Each piece of equipment is isolated by an optocoupler (or photocoupler), which provides isolation and avoids ground loop problems that are frustrating when processing a sound signal.

Materially, the connection is ensured by connectors designed around five-pin 180° DIN sockets (41524). As the protocol is serial, the materials are linked in a daisy chain. The cable length limit is 15 m, and they are protected and follow the pinout in Figure 5.4.

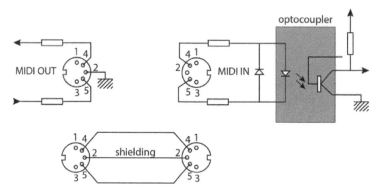

Figure 5.4. *Pinout of five-pin DIN sockets for MIDI*

Some manufacturers use TRS jack connectors with a diameter of 2.5 or 3.5 mm and provide adapters to remain compatible with the classic five-pin DIN.

---

2 Universal asynchronous receiver transmitter (UART) is a specialized electronic circuit, which transforms, using a clock and a shift register, parallel data into series data or vice versa.

NOTE.– After August 2018, the MMA recommends only the use of the 2.5-mm jack to avoid confusion with the audio connectors.

There are three possible MIDI sockets, and not all devices have them:

– MIDI IN: data input;

– MIDI OUT: data output;

– MIDI THRU: the information entering the IN passes directly to this output without being affected, in order to be able to connect other receiving equipment afterward.

Figure 5.5 shows the general principle and the connections of MIDI.

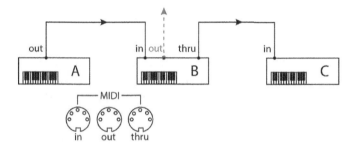

Figure 5.5. *Connections between equipment and MIDI connectors. In this configuration, equipment B and C are controlled from A (daisy-chain link)*

A link of the type shown in Figure 5.6, using only the IN and OUT ports, does not offer the same functionalities as those in Figure 5.5.

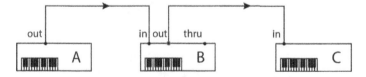

Figure 5.6. *In this configuration, equipment A controls B, B controls C, but A does not control C*

A MIDI link cannot exceed 15 m in length, otherwise there may be data corruption due to the attenuation imposed by the cable.

When transiting to THRU, the signal is re-amplified, as in Figure 5.7.

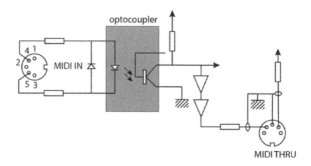

Figure 5.7. *General diagram of the IN to THRU link within the MIDI interface*

The hardware interface is not everything. It must be associated with software that ensures correct transmission of data, and this is the role of the MIDI protocol.

Today, many synthesizers transport MIDI information via a USB cable, which makes it possible to enrich the information transmitted and to add new independent controls. They are provided most of the time for use via a microcomputer (editor, configuration software, firmware update, specific software, etc.).

However, USB does not have the capacity to link several synthesizers or machines (groove box, drum machine, effects, etc.) together.

Since 2018, the MIDI Polyphonic Expression (MPE) standard allows users to modulate the sounds per note, whereas previously it was only possible per channel.

Figure 5.8. *Part of the back panel of a Moog Matriarch. Bottom, center, its MIDI connectors (IN, THRU, OUT) and its USB port can be seen. For a color version of this figure, see www.iste.co.uk/reveillac/synthesizers1.zip*

5.2.5. *The software*

It ensures the sending and receipt of MIDI data codified in a standard way, according to a particular syntax, which forms MIDI messages.

5.2.5.1. *Overview*

A message is made up of several codes, which represent a musical event such as pressing a key on a keyboard, using the modulation wheel or even a particular command such as a change of channel or a change of mode (omni, poly, mono, etc., see Table 5.4). In fact, a musical sequence is made up of a succession of MIDI messages that form a series of events made up of codes, in the form of bytes, which can be of two types, status bytes or data bytes.

When a status byte is not isolated, it characterizes the data bytes that follow it. This then forms a MIDI message, which reflects, for example, the action of the musician on the keyboard.

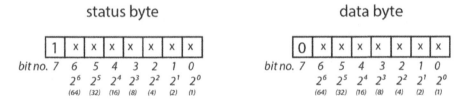

Figure 5.9. *Representation of a status byte and a data byte*

Since digital data are binary, the MIDI standard has defined that a byte beginning with 1 (most significant bit, MSB) represents a status and that a byte starting with 0 represents data.

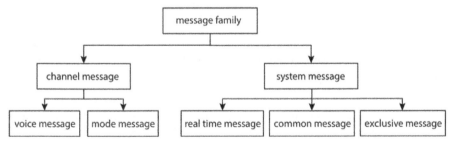

Figure 5.10. *Message families and their members*

The assembly of codes provides MIDI messages that can be divided into two large families: *Channel Messages* and *System Messages*.

The MIDI standard recognizes 16 channels. These channels address data that may be different for each of the connected devices, which can be configured to accept data from one or more channels. There are two reception modes, *Omni On* and *Omni Off*.

In the Omni Off mode, the hardware only receives on the channel it is set to; in Omni On mode, it receives on all channels.

NOTE.– To multiply the channels, it is possible to multiply the 16 output channels to several ports labeled with letters A, B, C and so on.

5.2.5.2. *Families*

The set of commands for the Channel Messages family is grouped into two members.

– *Voice Messages*: there are seven of them and note that the materials do not necessarily include the seven messages.

– *Mode Messages*: there are eight of them; the first four define the voice assignment of the hardware according to the MIDI channels and the other four have more specific functions.

Table 5.3 summarizes Voice Messages with their status and data bytes. nnnn represents the MIDI channel from 0 to 15 (channel numbers 1–16).

Table 5.4 summarizes Mode Messages in connection with the Control Change (CC).

– Table 5.5 presents the commands of the System Messages family, which are divided into two members.

– *System Real Time Messages*: there are eight system real-time messages (two of which are undefined) that take precedence over all other MIDI messages during a transmission. They apply to the entire MIDI system considered regardless of the channel.

– *System Common Messages*: there are eight of them (including two undefined ones), and they are intended for equipment connected to the MIDI system.

Command	Status byte (bin)	Data byte (bin)	Description
Note off	1000 nnnn	0kkkkkkk 0vvvvvvv	Determines if a note is no longer played kkkkkkk is the note – vvvvvvv is the velocity
Note on	1001 nnnn	0kkkkkkk 0vvvvvvv	Determines if a note has been played (pressing a key on the keyboard) kkkkkkk is the note – vvvvvvv is the velocity
Polyphonic key pressure or aftertouch	1010 nnnn	0kkkkkkk 0ppppppp	Message that characterizes the pressure exerted on a key Note: this function is not implemented on all synthesizers, expanders or various keyboards kkkkkkk is the note – ppppppp is the pressure
Control change	1011 nnnn	0ccccccc 0vvvvvvv	Modifies the parameters of the devices using the defined MIDI channel ccccccc is the control number – vvvvvvv is the control value
Program change	1100 nnnn	0ppppppp	Changes preset sounds across hardware using a defined MIDI channel ppppppp is the program number
Aftertouch or overall pressure or channel pressure	1101 nnnn	0vvvvvvv	Modifies the general setting of the dynamics of all the keys on the keyboard vvvvvvv is the pressure value
Pitch bender change	1110 nnnn	0nnnnnnn 0mmmmmmm	Determines the position of the pitch change wheel nnnnnnn is the LSB – mmmmmmm is the MSB

Table 5.3. *Voice Messages*

Command	Description
Mode 1	Omni on, poly on: voice messages are accepted regardless of their channel, and all voices produce the same sound (the number of voices is linked to the polyphony of the hardware)
Mode 2	Omni on, mono on: voice messages are accepted regardless of their channel and only one voice is played
Mode 3	Omni off, poly on: voice messages are accepted on the set MIDI channel, and all voices sound the same (the number of voices is related to the polyphony of the hardware)
Mode 4	Omni off, mono on: voice messages are accepted on the set MIDI channel, and each channel controls only one voice. The hardware splits into several monophonic ones (as many as there are voices available)
All sound off	Disables all voices on the defined channel. The sound is muted
Local control	Disconnects the keyboard from devices that have local control (local control off) or reconnects it (local control on)
Reset all controllers	Resets all controllers. The hardware returns to the state it was in when it was powered on
All notes off	Tells the hardware to mute all voices on the set channel

Table 5.4. *Mode Messages*

Member	Command	Description
Real time messages	MIDI timing clock	It is a signal that synchronizes several devices on the same tempo. The *MIDI timing clock* delivers 24 pulses for a quarter note[3] (PPQN)
	Start	Start signal transmitted for the master unit in the MIDI system (sequencer, drum machine, etc.). Its purpose is to synchronize the hardware from a musical sequence
	Stop	Unlike start, this signal stops all connected equipment
	Continue	Signal that is sent to all the equipment in the MIDI system, resuming from where they left off
	Active sensing	Signal that monitors the hardware. It controls all equipment and checks their MIDI connection. Every 300 ms, a message should arrive at the IN connection if no other message reaches it
	System reset	Resetting of the system's MIDI hardware. They must return to their state when they were powered on
Common messages	System exclusive	Allows manufacturers to create their own messages (like dump, patch parameters, etc.) and provides an additional MIDI specification message creation capability. It contains a manufacturer identification code assigned by the MMA or the AMEI on 1 or 3 bytes and universal exclusive messages
	MIDI time code	This is a real-time translation of the hours, minutes, seconds and frames per second of information of a Society of Motion Picture and Television Engineers (SMPTE) message. It is based on the SMPTE code structure and adapted to the splitting of the MIDI standard
	Song position pointer	2-byte precise marker (0–16,383) of a place in a sequence. This message is widely used by drum machines and sequencers
	Song select	Selects a sequence number (from 0 to 127) on a sequencer or drum machine
	Tune request	Asks the machine to agree on the frequency set by the manufacturer
	End of exclusive	This message marks the end of an exclusive message (EOX)

Table 5.5. *System Messages*

---

3 Here, "quarter note" means the value of a musical note in the same way as the "whole note", the "half note", the "eighth note", the "sixteenth note", etc.

5.2.6. *MIDI Control Change*

Table 5.3 refers to the so-called Control Change (CC) command. While CCs were already widely used when MIDI appeared, over time, with the improvement and evolution of all kinds of equipment, they have become essential to control more and more parameters.

We can divide them into three large groups:

– Continuous controllers that carry information on 128 values, from 0 to 127. They are located on slots 0–63.

– Discrete controllers that carry switch, on/off type information. A value equal to 0 or between 0 and 63 indicates that the function is stopped. A value equal to 127 or between 64 and 127 means that the function is active.

– The mode controllers that configure all the voices (*Channel Mode Messages*).

Table 5.6 summarizes the main MIDI CCs.

NOTE.– The hexadecimal values presented in the rest of this chapter will adopt the standard notation convention 0x to differentiate them from decimal values.

For example, 0x4A for 74 in decimal form or 0x7C for 124 in decimal form.

Since MIDI aims to guarantee perfect backward and forward compatibility, the heart of MIDI is designed not to evolve. Thus, current messages can also be specified by *Registered Parameter Numbers* (RPNs) present in the standard.

A standard MIDI message, as mentioned previously, can take values between 0 and 127, that is to say 128 available values, which is sometimes limited. RPNs were created to avoid this obstacle.

RPNs use 2 bytes, a most significant byte (MSB) and a least significant byte (LSB), which can operate on a range of 16,384 values. In return, each command parameter requires more data to be understood and processed.

Four messages make up an RPN parameter, the first two identify the parameter (MSB – CC 101 (0x65) and LSB – CC 100 (0x64)) and the second two, the value or variation to be assigned to this same parameter (MSB – CC 6 (0x06) and LSB – CC 38 (0x26).

Decimal value	Hexadecimal value	Description
0	0x00	Control bank selection
1	0x01	Modulation wheel
2	0x02	Breath controller
3	0x03	*Undefined (MSB)*
4	0x04	Foot controller
5	0x05	Portamento time
6	0x06	Data entry
7	0x07	Channel volume
8	0x08	Balance
9	0x09	*Undefined (MSB)*
10	0x0A	Pan
11	0x0B	Expression volume
12	0x0C	Effect 1 control
13	0x0D	Effect 2 control
14-15	0x0E-0x0F	*Undefined (MSB)*
16-19	0x10-0x13	General purpose controllers numbered 1–4
20-31	0x14-0x1F	*Undefined (MSB)*
32-63	0x20-0x3F	LSB for controllers 0–31
64	0x40	Hold footswitch 1 or Dampp (sustain) pedal – Type On/Off
65	0x41	Portamento footswitch – Type On/Off
66	0x42	Sostenuto footswitch – Type On/Off
67	0x43	Soft footswitch – Type On/Off
68	0x44	Legato footswitch – Type On/Off
69	0x45	Hold footswitch 2 (note hold time) – Type On/Off
70	0x46	Sound controller 1 (default: sound variation)
71	0x47	Sound controller 2 (default: timbre/harmonic intensity)
72	0x48	Sound controller 3 (default: release time)
73	0x49	Sound controller 4 (default: attack time)
74	0x4A	Sound controller 5 (default: brightness)
75	0x4B	Sound controller 6 (default: decay time)
76	0x4C	Sound controller 7 (default: vibrato rate)
77	0x4D	Sound controller 8 (default: vibrato depth)
78	0x4E	Sound controller 9 (default: vibrato delay)
79	0x4F	Sound controller 10
80-83	0x50-0x53	General purpose controllers numbered 5–8
84	0x54	Portamento control

Decimal value	Hexadecimal value	Description
85-87	0x55-0x57	*Undefined*
88	0x58	Possible extension of the range of velocity values
89-90	0x59-0x5A	*Undefined*
91	0x5B	Effects 1 depth (reverb level)
92	0x5C	Effects 2 depth (formerly tremolo depth)
93	0x5D	Effects 3 depth (chorus level, formerly chorus depth)
94	0x5E	Effects 4 depth (formerly detune depth)
95	0x5F	Effects 5 depth (formerly phaser depth)
96	0x60	Data increment
97	0x61	Data decrement
98	0x62	Non-Registered Parameter Number FINE (NRPN) LSB Value
99	0x63	Non-Registered Parameter Number COARSE (NRPN) MSB Value
100	0x64	Registered Parameter Number FINE (RPN) LSB value
101	0x65	Registered Parameter Number COARSE (RPN) MSB value
102-119	0x66-0x77	*Undefined*
Channel Mode Messages		
120	0x78	All sound off. When the message is received, all oscillators are muted, and their volume is set to zero as soon as possible
121	0x79	Reset track controllers. Resets all track controllers to their default values depending on the connected device (or software)
122	0x7A	Enables/disables local control. Disabling local control forces devices on the channel to ignore all non-MIDI data. Activation restores the normal functions of the controllers
123	0x7B	All notes off. Stops all notes by stopping all oscillators
124*	0x7C	Omni Mode Off
125*	0x7D	Omni Mode On
126*	0x7E	Mono Mode On (Polyphonic Off)
127*	0x7F	Poly Mode On (Mono Off)

* Controllers 124, 125, 126 and 127 manage four different modes (124 or 125) + (126 or 127):
– Poly (127): all notes played are encoded on a single MIDI channel;
– Mono (126): each note played is encoded on a different MIDI channel;
– Omni Off (124): selects one channel for sending and one for receiving MIDI data;
– Omni On (125): all information is received regardless of channel.

Table 5.6. *MIDI Control Change*

To give manufacturers more freedom, provision has been made for the use of non-registered parameters numbers (NRPN), which may be different for each MIDI-compatible device.

As with RPNs, NRPNs use data consisting of four messages (MSB – CC99, LSB – CC98, MSB – CC 6, LSB – CC38).

5.2.7. *Examples of MIDI transmission*

After presenting the hardware and software, we will discuss the structure of some MIDI actions and the corresponding messages.

The objective of this section, similar to that of this chapter, is not to present the MIDI standard in all its aspects but to simply provide an overview.

For those wishing to go further, consult the reference section of this book.

NOTE.– All bytes with an MSB (7th bit) equal to 0 are status bytes and those with an MSB equal to 1 are data bytes. The status byte determines the message type. The number of data bytes that follow depends on this type. There is a notion of current state; if we send data without repeating the status byte, we simply consider that the same status byte is implied.

5.2.7.1. *Note-on/note-off messages*

This type of message is generated when the user presses a key on the keyboard or applies a physical action on a MIDI controller to produce a sound. When the generator receives this message, the requested note is played.

Example: the user plays note A2 (the 2nd) on the keyboard:

```
10010000 00111001 01110111  or 0x90 0x39 0x77

10000000 00111001 01000000  or 0x80 0x39 0x40.
```

Table 5.7 details this content.

Byte (bin)	Command	Description
`10010000`	Note-on MIDI channel	Pressing a key on the keyboard Selected MIDI channel, from 1 to 16 In this case, channel 1
`00111001`	MIDI note number	Corresponds to a note, between 0 and 127 (from C-2 to G8 – Do-2 to Sol8) In this case, A2 (the 2nd)
`01110111`	Velocity	Corresponds to the speed of pressing the key, between 0 and 127 In this case, 119
`10000000`	Note-off	Pressing a key on the keyboard Selected MIDI channel, from 1 to 16 In this case, channel 1
`00111001`	MIDI note number	Corresponds to a note, between 0 and 127 (from C-2 to G8 – Do-2 to Sol8) In this case, A2 (the 2nd)
`01000000`	Velocity	Corresponds to the key release speed, between 0 and 127 In this case, 64

Table 5.7. *Details of MIDI messages when a user plays note A2 (the 2nd) on a keyboard*

5.2.7.2. Program change message

Initially, the designers of the MIDI standard imagined being able to change the sound of musical equipment such as an expander or synthesizer. As technology evolved, other types of hardware, such as sound effects racks or drum machines, also implemented MIDI, thereby involving a change of patch, memory, preset and so on.

Example: the user calls program No. 25:

`11001001 00011001` or `0xC9 0x19`.

Byte (bin)	Command	Description
`11001001`	Program change MIDI channel	Pressing a key on the keyboard Selected MIDI channel, from 1 to 16. In this case, channel 10
`00111001`	Program number	Corresponds to a program number, between 0 and 127 In this case, 25

Table 5.8. *Details of MIDI messages when a user calls program 25*

5.2.7.3. Reverb level change message (effect 1)

This message is sent when the user wants to modify the level of effect 1. By default, this effect is reverb.

Example: The user chooses a reverberation level equal to 50:

```
10110000 01011011 00110010 or 0xB0 0x5B 0x32.
```

Byte	Command	Description
10110000	Control Change MIDI channel	Pressing a key on the keyboard Selected MIDI channel, from 1 to 16 In this case, channel 1
01011011	Call effect	Corresponds to the CC code of effect 1 (default: reverb)
00110010	Effect level value	Matches a value between 0 and 127 In this case, 50

Table 5.9. Details of MIDI messages when a user wants a reverb level of 50

5.2.7.4. RPN message for coarse adjustment

This ok message is required in the General MIDI 1 standard (see section 5.2.9). There are two types of adjustment: fine adjustment (0x0001) and coarse adjustment (0x0002).

Example: the user coarsely adjusts channel 0 on an La (A) = 440 Hz:

```
10110000 01100101 00000000 or 0xB0 0x65 0x00
10110000 01100100 00000002 or 0xB0 0x64 0x02
10110000 00000006 01000000 or 0xB0 0x06 0x40
10110000 01100100 01111111 or 0xB0 0x64 0x7F
10110000 01100101 01111111 or 0xB0 0x65 0x7F.
```

Byte	Command	Description
10110000	Control Change MIDI channel	Changing hardware settings Selected MIDI channel, from 1 to 16 In this case, channel 1
01100101	MSB	RPN
00000000	1st byte	Coarse Channel Tuning
10110000	Control Change MIDI channel	Changing hardware settings Selected MIDI channel, from 1 to 16 In this case, channel 1
01100100	LSB	RPN
00000002	2nd byte	Coarse Channel Tuning The 2 bytes MSB and LSB form a value 0x0002 interpreted as the coarse tuning of the channel
10110000	Control Change MIDI channel	Changing hardware settings Selected MIDI channel, from 1 to 16 In this case, channel 1
00000110	Data entry	Ready to receive value
01000000	Adjustment value	La (A) 440 Hz
10110000	Control Change MIDI channel	Changing hardware settings Selected MIDI channel, from 1 to 16 In this case, channel 1
01100100	LSB	RPN
01111111	End	Closes the LSB of the RPN
10110000	Control Change MIDI channel	Changing hardware settings Selected MIDI channel, from 1 to 16 In this case, channel 1
01100101	MSB	RPN
01111111	End	Closes the MSB of the RPN

Table 5.10. *RPN message for coarse adjustment*

```
                                                                Date:
                MODEL           MIDI Implementation Chart       Version:
+---------------------+----------------+----------------+----------------+
!       Function...   ! Transmitted    ! Recognized     ! Remarks        !
!---------------------+----------------+----------------+----------------!
!Basic    Default     !                !                !                !
!Channel  Changed     !                !                !                !
!---------------------+----------------+----------------+----------------!
!         Default     !                !                !                !
!Mode     Messages    !                !                !                !
!         Altered     !                !                !                !
!---------------------+----------------+----------------+----------------!
!Note                 !                !                !                !
!Number : True Voice! !                !                !                !
!---------------------+----------------+----------------+----------------!
!Velocity Note ON     !                !                !                !
!         Note OFF    !                !                !                !
!---------------------+----------------+----------------+----------------!
!After    Key's       !                !                !                !
!Touch    Ch's        !                !                !                !
!---------------------+----------------+----------------+----------------!
!Pitch Bender         !                !                !                !
!---------------------+----------------+----------------+----------------!
!                     !                !                !                !
!                     !                !                !                !
! Control             !                !                !                !
!                     !                !                !                !
! Change              !                !                !                !
!                     !                !                !                !
!                     !                !                !                !
!---------------------+----------------+----------------+----------------!
!Prog                 !                !                !                !
!Change   True #      !                !                !                !
!---------------------+----------------+----------------+----------------!
!System Exclusive     !                !                !                !
!---------------------+----------------+----------------+----------------!
!System : Song Pos    !                !                !                !
!       : Song Sel    !                !                !                !
!Common : Tune        !                !                !                !
!---------------------+----------------+----------------+----------------!
!System      :Clock   !                !                !                !
!Real Time :Commands! !                !                !                !
!---------------------+----------------+----------------+----------------!
!Aux    :Local ON/OFF !                !                !                !
!       :All Notes OFF!                !                !                !
!Mes-   :Active Sense !                !                !                !
!sages: :Reset        !                !                !                !
!---------------------+----------------+----------------+----------------!
! Notes               !                !                !                !
!                     !                !                !                !
!                     !                !                !                !
!                     !                !                !                !
!---------------------+----------------+----------------+----------------!
     Mode 1 : OMNI ON,  POLY     Mode 2 : OMNI ON,  MONO      O : Yes
     Mode 3 : OMNI OFF, POLY     Mode 4 : OMNI OFF, MONO      X : No
```

Figure 5.11. *The reference model for the MIDI implementation chart*

5.2.8. *MIDI implementation chart*

To specify all the parameters processed by a machine, the MMA and the JSMC designed a document called the "MIDI Implementation Chart", which must accompany a device.

Its form and content are standardized.

The first column specifies the type of message, the second the transmission functions, the third the reception functions and the last column any remarks.

If a type of message is implemented on the machine, it is indicated by a dot, otherwise by a cross.

5.2.9. General MIDI standard

The General MIDI (GM) standard dates back to January 1991. The MMA and the JMSC validated this new standard in turn.

Why was a new standard created and what are its advantages? We will try to answer this question.

Previously, when a musician composed a MIDI sequence, it was adapted to their own set of MIDI hardware, so much so that when playing it on other devices, all the parameters for the new configuration had to be redefined, in particular the *Program Changes* with the risk of hearing a melody played with a clarinet sound when it was intended for an electric guitar. Other parameters also could not be taken into account such as the number of tracks, often depending on the power of the sequencer and so on.

The GM standard remedies these problems by standardizing hardware behavior by means of imposed parameters. However, there is an element that is difficult to manage, the timbre of each instrument, which will vary from one machine to another. This is directly linked to the sound synthesis technique chosen by the manufacturer (FM synthesis, granular synthesis, sampled sounds, etc.).

Below are the main constraints imposed by the specifications of the GM standard:

– the sounds are defined according to 16 families of eight instruments – the table of 128 instruments, thus defined, must be complete;

– the sound materials must be multitimbral with 16 polyphony voices for the instruments and eight voices for the percussion sounds;

– channel 10 is dedicated to percussion sounds – each note is assigned to a sound, and there must be at least 47 sounds;

– all 16 MIDI channels must be operational;

– the note, 3rd C (C3), must correspond to the number 60 of the MIDI note.

The following MIDI control events will be taken into account:

– speed and aftertouch on all voices;

– the referenced parameter controllers for pitchbend;

– for Channel Voice Messages via Control Change:

- 1 – modulation (vibrato, LFO),

- 7 – general volume,

- 10 – voice panorama,

- 11 – expression,

- 64 – sustain;

– for all Channel Mode Messages:

- 121 – reset all controllers,

- 123 – all notes off;

– default general volume is 90;

– the initial chord is A (A) at 440 Hz;

– a system exclusive message must place (or not) the equipment in GM mode;

– devices conforming to the GM standard bear the symbol in Figure 5.12.

Figure 5.12. *GM logo*

5.2.10. *The General MIDI 2 standard*

In March 1999, manufacturers extended the GM standard because many users complained that it was not efficient enough. This is how the General MIDI Level 2 (GM2) standard, validated by the MMA, was born.

Figure 5.13. *General MIDI 2 logo*

As a result, the acronym of the GM standard became GM1 (General MIDI level 1).

The GM2 standard adds 87 new instruments to the previous standard, giving a total of 215, the number of effects increased to 46 and the percussive sounds practically tripled (from 47 in GM1 to 133 in GM2). Channel 11 is used in parallel with channel 10 for the simultaneous management of drum kits.

The number of voices was extended to 32 (32 simultaneous notes). There are more voice messages per channel, especially for Control Change, as well as universal exclusive messages.

These are the main changes; other features were also added. For more information, consult the reference section of this book.

To learn more about the families of instruments and assignments of the GM1 and GM2 standard, the reader can consult Appendix 1.

5.2.11. *The GS format*

This format was introduced in 1991. It is similar in design to the GM format, but it is only used by instruments from the manufacturer Roland. It provides 16,384 sounds and can manage 128 kits or percussion *sets* assigned to channel 10. In reality, the GS format usually only has 10 kits, and only one kit is required.

Figure 5.14. *The logo for the MIDI Roland GS format*

The following Channel Voice Messages are recognized despite some adjustments:

– the polyphonic aftertouch and the pitchbend are considered;

– for the Control Change:

- bank select (sound bank choice),

- modulation,

- portamento time,

- data entry,

- volume (general),

- pan (voice panorama),

- expression,

- hold,

- sostenuto,

- soft,

- effect 1 (reverb),

- effect 3 (chorus),

- RPN (referenced parameter controllers),

- NRPN (unreferenced parameter controllers).

Known Channel Mode Messages include:

– all sounds off;

– reset all controllers.

Some standard exclusive messages are also implemented.

5.2.12. *The XG format*

Similar to the GS standard, the XG standard is also proprietary. It was created by the Japanese manufacturer Yamaha. It is also an extension of the GM standard with which it remains compatible. However, it offers many additional possibilities:

– polyphony on 32 simultaneous voices;

– number of sound banks greater than 100, each of them comprising 128 sounds;

– presence of additional channels to manage percussion sounds parallel to channel 10;

– additional internal graphic equalizer available;

– three sound effects (chorus, reverb and insertion effect);

– real-time modification of voices;

– consideration of external inputs (microphone, electric guitar, etc.);

– extension of voice messages per channel, especially for Control Change.

Figure 5.15. *Logo for the Yamaha MIDI XG format*

5.2.13. *MIDI file structure*

A MIDI file is organized as a succession of data blocks called *chunks*. These blocks have the structure shown in Table 5.11.

Type	Length	Data
4 bytes	4 bytes	Data bytes (according to the defined length)

Table 5.11. *Structure of a data block*

There are two types of blocks, the header block, *MThd chunk* and the track block, *MTrk chunk*.

5.2.13.1. *Header block*

The MThd chunk block defines the MIDI file header. A MIDI file contains only one block of this type.

The first 4 bytes represent the MThd string in ACSII code (4D 54 68 64).

The next 4 bytes specify that the data length of this block will be 6 bytes (00 00 00 06).

The last 6 bytes are used two by two. The first two indicate the format of the MIDI file, the next two the number of tracks and the last two the time resolution.

Type	Length	Data		
4 bytes	4 bytes	6 data bytes		
MThd (ASCII)	32 bits	16 bits (2 bytes)	16 bits (2 bytes)	16 bits (2 bytes)
0x4D 0x54 0x68 0x64	0x00 0x00 0x00 0x06	<format>	<tracks>	<division>

Table 5.12. *Standard format of an MThd chunk block*

The format can be type 0, 1 or 2:

– type 0 indicates that the file has only one MIDI data track;

– type 1 indicates that the file contains several tracks, which will be playable simultaneously – each of the tracks contains information dedicated to a MIDI channel;

– type 2 indicates that the file contains *patterns* – each track is played independently of the others.

The number of tracks is always equal to 1 if the chosen format is 0.

The temporal resolution is expressed in *ticks* per quarter note or in delta-time units per SMPTE frame.

For a resolution in pulses, the MSB of the 5th byte is equal to 0 and the 15 others give the number.

	5th byte		6th byte
Bit	7	6-0	7-0
Division	0	Ticks per quarter note	

Table 5.13. *Tick resolution*

For a resolution in time intervals, the MSB of the 5th byte is equal to 1, bits from 6 to 0 of the first byte represent the number of frames per second (expressed as a negative number: –24, –25, –29 or –30 frames per second) and bits from 7 to 0 of the second byte represent the number of delta-time units per SMPTE.

	5the byte		6th byte
Bit	7	6-0	7-0
Division	1	Frames/second	Ticks/frame

Table 5.14. *Resolution in time intervals*

5.2.13.2. *Track block*

This block contains all the MIDI data for each track. It consists of a sequence of events associated with a time interval (delta time).

If the chosen format is type 0, the track block will contain all the notes and the information relating to the tempo.

If the format is type 1, the first track block is specific, and it is called the *tempo map*. This is where the elements will be defined:

– time signature: MIDI sequence time parameters;

– set tempo: setting the tempo value;

– sequence/track name: MIDI track or sequence name.

If the format is type 2, each track block represents an independent MIDI sequence.

The general format of the track block is shown in Table 5.15.

Type	Length	Data
4 bytes	4 bytes	Data bytes
MTrk (ascii)	32 bits	According to the defined length
0x4D 0x54 0x72 0x6B	<length>	<time intervals> <events>…

Table 5.15. *Standard track block format*

Time intervals are used to define the temporal aspect of a MIDI sequence. The data that constitute a MIDI sequence will integrate time intervals that specify the time elapsed between two MIDI events. Events accept the *running status*[4]. To limit the size of the files, the time intervals undergo compression via a representation of variable size expressed using 1, 2, 3 or 4 bytes.

To recognize the bytes that constitute a time interval, we act on the MSB. If it is equal to 0, it is the last byte.

1 byte: they have a value between 0 and 127, 00000000 to 01111111.

After compression, the values remain the same, the MSB is equal to 0, so it is the only byte in the time interval.

---

4 Since the structure of MIDI data is often repetitive, manufacturers use a technique called the *running status*, which reduces the size of the data and mitigates certain transmission delay problems. During the transmission of consecutive messages with the same status carried by the same channel, the status byte is transmitted only once.

2 bytes: they have a value between 128 and 16,383, 00000000 10000000 to 00111111 11111111.

After compression, the values become:

10000001 00000000 to 11111111 01111111.

Bits 7–13 have been shifted one bit to the left, bit 7 set to 0 and bit 15 set to 1.

3 bytes: they have a value between 16,384 and 2,097,151, 00000000 10000000 00000000 to 00011111 11111111 11111111.

After compression, the values become:

10000001 10000000 00000000 to 11111111 11111111 01111111.

Bits 14–20 have been shifted two bits to the left, bits 7–13 one bit to the left, bit 7 set to 0 and bits 15 and 23 set to 1.

4 bytes: they have a value between 2,097,152 and 268,435,455, 00000000 00100000 00000000 00000000 to 00001111 11111111 11111111 11111111.

After compression, the values become:

10000001 10000000 10000000 00000000 to 11111111 11111111 11111111 01111111.

We shifted bits 21–27 three bits to the left, bits 14–20 two bits to the left and bits 7–13 one bit to the left, set bit 7 to 0 and positioned bits 15, 23 and 31 to 1.

There are three types of MIDI events contained in a track block, channel messages, (see section 5.2.5.2), system exclusive (*sysex event*) events and meta-events.

A system exclusive event or a meta-event cancels the running status.

5.2.13.3. *System exclusive events*

In general, they have variable lengths, starting with F0 and ending with an EOX (end of exclusive message) F7.

Start	Length	Bytes to transmit	End
0xF0	0xaa	0xbb 0xbb 0xbb ...	F7

Table 5.16. *Standard format of an exclusive event*

There are two special cases:

– A system exclusive event split into several subsets separated by time intervals. In this case, only the first starts with F0, the following always start with F7.

Start	Length	Bytes to transmit
0xF0	0xaa	0xbb 0xbb 0xbb ...

Time interval
0xcc ... 0xcc

Start	Length	Bytes to transmit
0xF7	0xaa	0xbb 0xbb 0xbb ...

Time interval
0xcc ... 0xcc

-
-
-
-

Start	Length	Bytes to transmit
0xF7	0xaa	0xbb 0xbb 0xbb ...

Table 5.17. *A complex system exclusive event*

– A system exclusive event that starts with F7 is used for the transmission of specific messages.

Start	Length	Bytes to transmit	End
0xF7	0xaa	0xbb 0xbb 0xbb ...	F7

Table 5.18. *A system exclusive event*

5.2.13.4. Meta-events

There are 16 meta-events:

– sequence number;

– text event;

– copyright notice;

– sequence/track name;

– instrument name;

– lyric;

– marker;

– cue point;

– program name;

– device name;

– end of track;

– set tempo;

– SMPTE offset;

– time signature;

– key signature;

– sequencer specific meta-event.

1) Sequence number:

- Optional event that must be used before any time interval or other events. It specifies the number of the sequence on 2 bytes or the pattern number in format 2.

- Format: `0xFF 0x00 0x02 0xnn 0xnn`.

2) Text:

- Variable length description text encoded in ASCII.

- Format: `0xFF 0x01 0xlen 0xtext` (string length, ASCII description text).

3) Copyright:

- Defines the copyright, author's name, year, © and so on, in ASCII, according to a defined length.

- Format: `0xFF 0x02 0xlen 0xtext` (string length, ASCII text).

4) Track or sequence name:

- Sequence name in format 0 or track name in format 1.

- Format: 0xFF 0x03 0xlen 0xtext (string length, ASCII text).

5) Instrument name:

- Text indicating the name of the MIDI instrument used.

- Format: 0xFF 0x04 0xlen 0xtext (string length, ASCII text).

6) Lyric:

- Song lyrics often broken down into syllables.

- Format: 0xFF 0x05 0xlen 0xtext (string length, ASCII text).

7) Marker:

- Text type marker associated with a specific time. It usually indicates a specific element like a chorus or a verse.

- Format: 0xFF 0x06 0xlen 0xtext (string length, ASCII text).

8) Cue point:

- Description in text form of an event placed at a specific time.

- Format: 0xFF 0x07 0xlen 0xtext (string length, ASCII text).

9) Program name:

- Name of the program set to play the track block. It can be different from the track name or the sequence name.

- Format: 0xFF 0x08 0xlen 0xtext (string length, ASCII text).

10) Device name (MIDI port name):

- Name of the MIDI port to which the track will be routed. When a port name has been defined, all events are directed to it. It is generally used in format 1 to bind a track to a port.

- Format: 0xFF 0x09 0xlen 0xtext (string length, ASCII text).

11) End of track:

- Mandatory event. It indicates the last event in each track block. It is unique per track block.

- Format: 0xFF 0x2F 0x00.

12) Set tempo:

- Variation in tempo indicated in microseconds for a quarter note for 3 bytes. By default, the tempo is 120 beats per minute (BPM).

- Format: `0xFF 0x51 0x03 0xtt 0xtt 0xtt`.

13) SMPTE offset:

- Starting point of the SMPTE code in hours, minutes, seconds, images per second (frames) and hundredths of an image per second (subframes).

- Format: `0xFF 0x54 0x05 0xhr 0xmn 0xse 0xfr 0xff`.

14) Time signature:

- Measurement of the musical sequence. Four bytes make up the numerator, denominator, number of MIDI clock messages (24 per quarter note) per metronome beat and number of notes in 24 MIDI clock beats (number of 32nd notes or 32nd notes per quarter note, respectively).

- The numerator nn is represented in the classic way.

- The denominator dd represents the power to which the number 2 is raised, which will characterize the measurement: $2^0 = 1$ (1 whole note to a whole note), $2^1 = 2$ (2 half notes to a whole note), $2^2 = 4$ (4 quarter notes to a whole note), $2^3 = 8$ (8 eighth notes to a whole note) and so on.

- Format: `0xFF 0x58 0x04 0xnn 0xdd 0xcc 0xbb`.

15) Key signature:

- Key signature, that is, the key and the mode or type of scale.

- The key is defined by `sf` with 0 for C major, positive values from 1 to 7 for sharps, negative values from −1 to −7 for flats.

- The mode is equal to 0 if the scale is major and 1 if the scale is minor.

- Format: `0xFF 0x59 0x02 0xsf 0xmi`.

- Example: `0xFF 0x59 0x02 0x07 0x00` – Do# major (Cmaj#).

16) Specific event:

- Manufacturer identification associated with specific messages dedicated to a given instrument.

- Format: `0xFF 0x7F 0xlen 0xdata` (string length, specific data).

5.2.14. *An example of a MIDI file*

You will find, in Table 5.19, all the data constituting a simple MIDI file, including the header block, the track block for the tempo and the track block describing a musical sequence on MIDI channel No. 1.

The hexadecimal codes making up the file are detailed and commented on.

Hexadecimal code	Comments
Chunk header	
0x4D 0x54 0x68 0x64	ASCII string: MThd
0x00 0x00 0x00 0x06	Block data length: 6
0x00 0x01	MIDI Format: 1
0x00 0x05	Number of tracks: 5
0x01 0x80	384 ticks (per quarter note)
Track block – Chunk track – (tempo track)	
0x4D 0x54 0x72 0x6B	ASCII string: MTrk
0x00 0x00 0x00 0x19	Block data length: 25
0x00	Delta times: 0
0xFF 0x51 0x03	Event: tempo
0x0B 0x71 0xB0	Tempo value: 750,000 µs/quarter note, i.e.: (60 s/0.750 s = 80 quarter notes/min or 80 BPM)
0x00	Delta times: 0
0xFF 0x58 0x04	Event: measurement
0x04	Numerator of the measurement: 4
0x02	Denominator of the measurement: $2^2 = 4$
0x18	Number of ticks of the metronome (midi clocks): 24 in a quarter note
0x08	Number of 32nd notes in a quarter note: 8
0x00	Delta times: 0
0xFF 0x59 0x02	Tone
0x00	0 (no alteration) => do
0x00	0 => major (1 => minor)
0x00	Delta time: 0
0xFF 0x2F 0x00	End of track
Track block No. 2 – Chunk track 2 (channel 1)	
0x4D 0x54 0x72 0x6B	ASCII string: MTrk
0x00 0x00 0x02 0x47	Block data length: 583
0x00 0xFF 0x03 0x09	Track name
0x53 0x74 0x65 0x65 0x6C 0x47 0x74 0x72 0x00	ASCII string: steelgtr

0x00	Delta times: 0
0xB1	Control Change/channel 2
0x07 0x64	Main volume: 100
0x00	Delta times: 0
0x0A 0x2C	Pan (panorama): 44
0x00	Delta times: 0
0x5B 0x37	Effects 1 depth (reverb): 55
0x00	Delta times: 0
0x5D 0x14	Effects 3 depth (chorus): 20
0x8C 0x04	Delta times: 1540
0x90	Note-on/channel 1
0x40 0x73	Note: E3 – velocity: 115
0x00	Delta times: 0
0x44 0x5B	Note: G#3 – velocity: 91
0x00	Delta times: 0
0x47 0x63	Note: B3 – velocity: 99
0x00	Delta times: 0
0x4C 0x5E	Note: E4 – velocity: 94
0x70	Delta times: 112
0x80	Note-off/channel 1
0x44 0x00	Note: G#3 – velocity: 0
0x00	Delta times: 0
0x47 0x00	Note: B3 – velocity: 0
0x08	Delta times: 8
0x40 0x00	Note: E3 – velocity: 0
0x00	Delta times: 0
0x4C 0x00	Note: E4 – velocity: 0
0x81 0x24	Delta times: 100
0x90	Note-on/channel 1
0x44 0x5D	Note: G#3 – velocity: 93
0x00	Delta times: 0
0x47 0x61	Note: B3 – velocity: 97
0x04	Delta times: 4
0x40 0x66	Note: E3 – velocity: 102
0x00	Delta times: 0
0x4C 0x59	Note: E4 – velocity: 89
0x2C	Delta times: 44
0x80	Note-off/channel: 1
0x44 0x00	Note: G#3 – velocity: 0
0x00	Delta times: 0

0x47 0x00	Note: B3 – velocity: 0
0x00	Delta times: 0
0x40 0x00	Note: E3 – velocity: 0
0x00	Delta times: 0
0x4C 0x00	Note: E4 – velocity: 0
0x34	Delta times: 52
–	–
–	–
–	–
0x00	Delta times: 0
0xFF 0x2F 0x00	End of track

Table 5.19. *An example of a MIDI file*

5.3. MIDI CV/Gate converters

If you have a vintage synthesizer, the temptation is great to be able to control it MIDI or to retrieve CV/Gate information in MIDI format.

For many years, there have been gateways or converters that perform these operations. Many even had a USB port allowing you to go even further in the exchange and conversion of data and analog signals.

Figure 5.16. *Doepfer MCV4 MIDI to CV/Gate converter*

With the arrival of modular synthesizers, many modules also provided these conversions to ensure communication with other equipment.

You will find some interfaces in Table 5.20. The list is far from exhaustive.

Manufacturer	Model	Characteristics
ACL	Mini MIDI	USB MIDI to CV – Eurorack Outputs: Pitch-CV, MIDI CC, MIDI Clock 2 USB ports, host and MIDI
Behringer	CM1A	MIDI to CV/Gate – Eurorack 2 CV outputs, 2 Gate outputs 1 USB-MIDI IN-THRU port
Doepfer	MCV4	MIDI to CV/Gate 4 CV outputs, 1 Gate output MIDI IN-THRU
Doepfer	A190-3	USB/MIDI to CV/Gate - Eurorack 4 CV outputs 1 Gate output MIDI IN – USB port
Doepfer	A192-2	Dual CV/Gate to MIDI/USB – Eurorack 2 independent interfaces 4 inputs per interface: Gate, CVN, CVV, CVC MIDI IN-OUT – USB port
Doepfer	Dark Link	USB/MIDI to CV/Gate 4 CV outputs, 1 Gate output Glide function MIDI In – USB port – 5 V and 12 V Gate compatible
Kenton	PRO CV to MIDI	CV/Gate to MIDI 2 programmable CV inputs 2 aux inputs and Gate Hz/Volt scale, 1 V/oct, 1.2 V/Oct 32 memory slots MIDI IN-OUT
MST	MIDI to CV	MIDI to CV – Eurorack MIDI IN-OUT – USB port Clock input/output 1 CV output, 1 Gate output, auxiliary CV

Table 5.20. *Some MIDI –> CV/Gate or CV/Gate –> MIDI converters*

NOTE.– In Appendix 2, you will find other useful materials for certain hardware configurations using MIDI.

Conclusion

After reading Volume 1, you should have discovered or reinforced your knowledge of sound and subtractive synthesis.

The fundamentals related to acoustics, propagation and listening to sound in their historical context are in Chapter 1.

The journey continued through the many twists and turns attached to sound synthesis techniques that have emerged since the beginning of the electronic creation of sounds in the middle of the 20th century, until today.

Armed with all this information, you have been able to take a glimpse at the tools, functions and processing that subtractive synthesizers must implement to clone the range, the envelope, the timbre and even the format of a musical instrument, not to mention the possibilities of imitating the noises and sounds present in the world around us. Let us not forget sound design and the creation of new sounds for the media, cinema, television, the advertising industry, art and so on.

Since the birth of the first artificial sound-creating devices, such as the church organ, the telharmonium, the theremin and the ondioline, technology has constantly evolved; hundreds of machines have appeared. To familiarize you with subtractive sound synthesis, several of them have been selected for this book, from the past five decades. Hardware (ARP 2600, Minimoog, MatrixBrute from Arturia, etc.) and software (VCV Rack, Max/MSP, etc.) are presented, and they prefigure the work and exercises detailed in Volume 2.

Chapter 5 is devoted to the CV/Gate and MIDI standards, essentials that cannot be overlooked when dealing with electronic music and sound synthesis.

Acoustics, particularly in simulation, is progressing daily, and computing power is evolving periodically, as Gordon E. Moore showed so well in 1965, and immense progress has been made in only a few decades. As for electronics, the transition from analog to digital has ignited the creativity of the main players in this field: inventors, manufacturers and developers are constantly innovating. Not a week goes by without a new device showing up, constantly shaking up the musical universe.

In this cosmopolitan and technological present, trends appear, and the fundamentals are ever present; vintage synthesizers have never been so popular with professionals and amateurs, and the electronic music of the 21st century is here to testify.

Several manufacturers are pushing back the limits of synthesis by offering compact machines with very advanced characteristics, while others are revisiting the synthesizers of the 1970s and modernizing them (MIDI, USB, editor, software controls, etc.) at much lower cost than those of the time, thus making them very accessible. With the soaring spread of microcomputers, synthesizers and virtual effects invading macOS, Microsoft Windows and even Linux, there are no limits, and it has become challenging to navigate.

The last few years have seen the arrival of many modular synthesizers, and manufacturers are having the time of their lives. Thousands of processing and synthesis modules are available on the market. As if that were not enough, software clones of these same modules are also available, allowing for the assembly and implementation of hybrid machines.

Either way, no one can escape digital. Archiving, editing, control, exploitation or creation of synthesizers require the implementation of information technologies (IT). More than ever, the microcomputer has become an instrument in its own right.

After this brief overview of the evolution of sound synthesis, it is time to tackle Volume 2 of this book. You can practice and apply the concepts, theories and principles developed in the previous five chapters.

Appendix 1

General MIDI 1 and 2 Instruments

For both General MIDI standards, each manufacturer of electronic instruments must ensure that their sounds provide an acceptable representation of the sound reproduction corresponding to the defined instrument.

Instrument names indicate what type of sound will be heard when that instrument number (MIDI program change or "PC#") is selected on a GM synthesizer.

These sounds are the same for all MIDI channels except channel 10 or 11 (GM2), which only has percussion sounds.

A1.1. List of General MIDI instruments

In Table A1.1, a list of all MIDI instruments of the General MIDI standard (GM1) with their patch number can be found. They are classified by instrumental family: piano, chromatic percussion, organ, etc.

	Piano		Reed
1	Acoustic grand piano	65	Soprano sax
2	Bright acoustic piano	66	Alto sax
3	Electric grand piano	67	Tenor sax
4	Honky-tonk piano	68	Baritone sax
5	Electric piano 1	69	Oboe
6	Electric piano 2	70	English horn
7	Harpsichord	71	Bassoon
8	Clavinet	72	Clarinet

	Chromatic Percussion		Pipe
9	Celesta	73	Piccolo
10	Glockenspiel	74	Flute
11	Music box	75	Recorder
12	Vibraphone	76	Pan flute
13	Marimba	77	Blown bottle
14	Xylophone	78	Shakuhachi
15	Tubular bells	79	Whistle
16	Dulcimer	80	Ocarina
	Organ		Synth Lead
17	Drawbar organ	81	Lead 1 (square)
18	Percussive organ	82	Lead 2 (sawtooth)
19	Rock organ	83	Lead 3 (calliope)
20	Church organ	84	Lead 4 (chiff)
21	Reed organ	85	Lead 5 (charang)
22	Accordion	86	Lead 6 (voice)
23	Harmonica	87	Lead 7 (fifths)
24	Tango accordion	88	Lead 8 (bass + lead)
	Guitar		Synth Pad
25	Acoustic guitar (nylon)	89	Pad 1 (new age)
26	Acoustic guitar (steel)	90	Pad 2 (warm)
27	Electric guitar (jazz)	91	Pad 3 (polysynth)
28	Electric guitar (clean)	92	Pad 4 (choir)
29	Electric guitar (muted)	93	Pad 5 (bowed)
30	Overdriven guitar	94	Pad 6 (metallic)
31	Distortion guitar	95	Pad 7 (halo)
32	Guitar harmonics	96	Pad 8 (sweep)
	Bass		Synth Effects
33	Acoustic bass	97	FX 1 (rain)
34	Electric bass (finger)	98	FX 2 (soundtrack)
35	Electric bass (pick)	99	FX 3 (crystal)
36	Fretless bass	100	FX 4 (atmosphere)
37	Slap bass 1	101	FX 5 (brightness)
38	Slap bass 2	102	FX 6 (goblins)
39	Synth bass 1	103	FX 7 (echoes)
40	Synth bass 2	104	FX 8 (sci-fi)

	Strings		Ethnic
41	Violin	105	Sitar
42	Viola	106	Banjo
43	Cello	107	Shamisen
44	Contrabass	108	Koto
45	Tremolo strings	109	Kalimba
46	Pizzicato strings	110	Bag pipe
47	Orchestral harp	111	Fiddle
48	Timpani	112	Shanai
	Strings (continued)		Percussive
49	String ensemble 1	113	Tinkle bell
50	String ensemble 2	114	Agogo
51	Synth strings 1	115	Steel drums
52	Synth strings 2	116	Woodblock
53	Choir aahs	117	Taiko drum
54	Voice oohs	118	Melodic tom
55	Synth voice	119	Synth drum
56	Orchestra hit	120	Reverse cymbal
	Brass		Sound effects
57	Trumpet	121	Guitar fret noise
58	Trombone	122	Breath noise
59	Tuba	123	Seashore
60	Muted trumpet	124	Bird tweet
61	French horn	125	Telephone ring
62	Brass section	126	Helicopter
63	Synth brass 1	127	Applause
64	Synth brass 2	128	Gunshot

Table A1.1. *General MIDI instruments*

A1.2. Drum note numbers (channel 10)

Unlike a melodic instrument, a percussion instrument always emits the same sound with a fixed pitch.

In General MIDI, each percussion instrument is associated with a different key on the keyboard (note). For some tuned percussion instruments, such as timpani, there will be several different keys.

In Table A1.2, the notes are those of a piano keyboard. Key 69 corresponds to A 440 (A4) of the fourth octave and key 27 to D# (D#1) of the first octave.

Note (piano)	Instrument name	Note (piano)	Instrument name
35	Acoustic bass drum	59	Ride cymbal 2
36	Bass drum 1	60	Hi bongo
37	Side stick	61	Low bongo
38	Acoustic snare	62	Mute hi conga
39	Hand clap	63	Open hi conga
40	Electric snare	64	low conga
41	Low floor tom	65	High timbale
42	Closed hi-hat	66	Low timbale
43	High floor tom	67	High agogo
44	Pedal hi-hat	68	Low agogo
45	Low tom	69	Cabasa
46	Open hi-hat	70	Maracas
47	Low-mid tom	71	Short whistle
48	Hi-mid tom	72	Long whistle
49	Crash cymbal 1	73	Short guiro
50	High tom	74	Long guiro
51	Ride cymbal 1	75	Claves
52	Chinese cymbal	76	Hi woodblock
53	Ride bell	77	Low woodblock
54	Tambourine	78	Mute cuica
55	Splash cymbal	79	Open cuica
56	Cowbell	80	Mute triangle
57	Crash cymbal 2	81	Open triangle
58	Vibraslap	–	–

Table A1.2. *List of percussions (channel 10)*

A1.3. Lists of General MIDI 2 instruments

Most General MIDI synthesizers are limited to the 128 instruments in bank 0.

General MIDI 2 compatible synthesizers can access all 256 instruments by setting Control Change CC#32 to 121 and using Control Change CC#0 to select the bank before calling a Program Change function.

Piano		
Patch number	**Bank number**	**Instrument name**
1	0	Acoustic grand piano
	1	Wet acoustic grand
	2	Dry acoustic grand
2	0	Bright acoustic piano
	1	Wet bright acoustic
3	0	Electric grand piano
	1	Wet electric grand
4	0	Honky-tonk piano
	1	Wet honky-tonk
5	0	Rhodes piano
	1	Detuned electric piano 1
	2	Electric piano 1 variation
	3	60's electric piano
6	0	Chorused electric piano
	1	Detuned electric piano 2
	2	Electric piano 2 variation
	3	Electric piano legend
	4	Electric piano phase
7	0	Harpsichord
	1	Coupled harpsichord
	2	Wet harpsichord
	3	Open harpsichord
8	0	Clavinet
	1	Pulse clavinet

Chromatic percussion		
Patch number	Bank number	Instrument name
9	0	Celesta
10	0	Glockenspiel
11	0	Music box
12	0	Vibraphone
	1	Wet vibraphone
13	0	Marimba
	1	Wet marimba
14	0	Xylophone
15	0	Tubular bell
	1	Church bell
	2	Carillon
16	0	Santur
Organ		
Patch number	Bank number	Instrument name
17	0	Hammond organ
	1	Detuned organ 1
	2	60's organ 1
	3	Organ 4
18	0	Percussive organ
	1	Detuned organ 2
	2	Organ 5
19	0	Rock organ
20	0	Church organ 1
	1	Church organ 2
	2	Church organ 3
21	0	Reed organ
	1	Puff organ
22	0	French accordion
	1	Italian accordion
23	0	Harmonica
24	0	Bandoneon

Guitar		
Patch number	**Bank number**	**Instrument name**
25	0	Nylon-string guitar
	1	Ukulele
	2	Open nylon guitar
	3	Nylon guitar 2
26	0	Steel-string guitar
	1	12-string guitar
	2	Mandolin
	3	Steel + body
27	0	Jazz guitar
	1	Hawaiian guitar
28	0	Clean electric guitar
	1	Chorus guitar
	2	Mid tone guitar
29	0	Muted electric guitar
	1	Funk guitar
	2	Funk guitar 2
	3	Jazz man
30	0	Overdriven guitar
	1	Guitar pinch
31	0	Distortion guitar
	1	Feedback guitar
	2	Distortion rtm guitar
32	0	Guitar harmonics
	1	Guitar feedback
Bass		
Patch number	**Bank number**	**Instrument name**
33	0	Acoustic bass
34	0	Fingered bass
	1	Finger slap
35	0	Picked bass
36	0	Fretless bass

Bass		
Patch number	**Bank number**	**Instrument name**
37	0	Slap bass 1
38	0	Slap bass 2
39	0	Synth bass 1
	1	Synth bass 101
	2	Synth bass 3
	3	Clavi bass
	4	Hammer
40	0	Synth bass 2
	1	Synth bass 4
	2	Rubber bass
	3	Attack pulse
Solo orchestra		
Patch number	**Bank number**	**Instrument name**
41	0	Violin
	1	Slow violin
42	0	Viola
43	0	Cello
44	0	Contrabass
45	0	Tremolo strings
46	0	Pizzicato strings
47	0	Harp
	1	Yangqin
48	0	Timpani
Ensemble orchestra		
Patch number	**Bank number**	**Instrument name**
49	0	String ensemble
	1	Orchestra strings
	2	60's strings
50	0	Slow string ensemble
51	0	Synth strings 1
	1	Synth strings 3

\multicolumn{3}{c}{Ensemble orchestra}		
Patch number	Bank number	Instrument name
52	0	Synth strings 2
53	0	Choir aahs
	1	Choir aahs 2
54	0	Voice oohs
	1	Humming
55	0	Synth voice
	1	Analog voice
56	0	Orchestra hit
	1	Bass hit
	2	6th hit
	3	Euro hit
\multicolumn{3}{c}{Brass}		
Patch number	Bank number	Instrument name
57	0	Trumpet
	1	Dark trumpet
58	0	Trombone
	1	Trombone 2
	2	Bright trombone
59	0	Tuba
60	0	Muted trumpet
	1	Muted trumpet 2
61	0	French horns
	1	French horn 2
62	0	Brass section 1
	1	Brass section 2
63	0	Synth brass 1
	1	Synth brass 3
	2	Analog brass 1
	3	Jump brass
64	0	Synth brass 2
	1	Synth brass 4
	2	Analog brass 2

Reed		
Patch number	Bank number	Instrument name
65	0	Soprano sax
66	0	Alto sax
67	0	Tenor sax
68	0	Baritone sax
69	0	Oboe
70	0	English horn
71	0	Bassoon
72	0	Clarinet
Wind		
Patch number	Bank number	Instrument name
73	0	Piccolo
74	0	Flute
75	0	Recorder
76	0	Pan flute
77	0	Bottle blow
78	0	Shakuhachi
79	0	Whistle
80	0	Ocarina
Synth lead		
Patch number	Bank number	Instrument name
81	0	Square lead
81	1	Square wave
81	2	Sine wave
82	0	Saw lead
82	1	Saw wave
82	2	Doctor solo
82	3	Natural lead
82	4	Sequenced saw
83	0	Synth calliope
84	0	Chiffer lead

Appendix 1   227

Synth lead		
Patch number	Bank number	Instrument name
85	0	Charang
	1	Wire lead
86	0	Solo synth vox
87	0	5th saw wave
88	0	Bass & lead
	1	Delayed lead
Synth pad		
Patch number	Bank number	Instrument name
89	0	Fantasia pad
90	0	Warm pad
	1	Sine pad
91	0	Polysynth pad
92	0	Space voice pad
	1	Itopia
93	0	Bowed glass pad
94	0	Metal pad
95	0	Halo pad
96	0	Sweep pad
Synth sound FX		
Patch number	Bank number	Instrument name
97	0	Ice rain
98	0	Soundtrack
99	0	Crystal
	1	Synth mallet
100	0	Atmosphere
101	0	Brightness
102	0	Goblin
103	0	Echo drops
	1	Echo bell
	2	Echo pan
104	0	Star theme

Ethnic		
Patch number	Bank number	Instrument name
105	0	Sitar
	1	Sitar 2
106	0	Banjo
107	0	Shamisen
108	0	Koto
	1	Taisho koto
109	0	Kalimba
110	0	Bagpipe
111	0	Fiddle
112	0	Shanai

Percussive		
Patch number	Bank number	Instrument name
113	0	Tinkle bell
114	0	Agogo
115	0	Steel drums
116	0	Woodblock
	1	Castanets
117	0	Taiko
	1	Concert bass drum
118	0	Melodic tom 1
	1	Melodic tom 2
119	0	Synth drum
	1	Analog tom
	2	Electric percussion
120	0	Reverse cymbal

Sound effect		
Patch number	Bank number	Instrument name
121	0	Guitar fret noise
	1	Guitar cut noise
	2	String slap

Appendix 1 229

Sound effect		
Patch number	Bank number	Instrument name
122	0	Breath noise
	1	Flute key click
123	0	Seashore
	1	Rain
	2	Thunder
	3	Wind
	4	Stream
	5	Bubble
124	0	Bird tweet
	1	Dog
	2	Horse gallop
	3	Bird 2
125	0	Telephone 1
	1	Telephone 2
	2	Door creaking
	3	Door closing
	4	Scratch
	5	Wind chimes
126	0	Helicopter
	1	Car engine
	2	Car stop
	3	Car pass
	4	Car crash
	5	Siren
	6	Train
	7	Jetplane
	8	Starship
	9	Burst noise

Sound effect		
Patch number	**Bank number**	**Instrument name**
127	0	Applause
	1	Laughing
	2	Screaming
	3	Punch
	4	Heartbeat
	5	Footsteps
128	0	Gunshot
	1	Machine gun
	2	Laser gun
	3	Explosion

Table A1.3. *General MIDI 2 instruments*

A1.4. GM2 drum note numbers (channel 10 and channel 11)

Access to different drum kits available in GM2 is via a bank change command (Control Change CC#0) with the program number (Program Change) of the desired drum kit. In GM2 mode, nine drum kits are available.

Program number	Kit
0	Standard set
1	Room set
2	Power set
3	Electric set
4	Analog set
5	Jazz set
6	Brush set
7	Orchestra set
8	SFX set

Table A1.4. *GM2 percussion kits*

Note (piano)	Standard set	Room set	Power set	Electric set	Analog set
D#0	High Q	High Q	High Q	High Q	High Q
E0	Slap	Slap	Slap	Slap	Slap
F0	Scratch push	Scratch push	Scratch push	Scratch push	Scratch push
F#0	Scratch pull	Scratch pull	Scratch pull	Scratch pull	Scratch pull
G0	Sticks	Sticks	Sticks	Sticks	Sticks
G#0	Square click	Square click	Square click	Square click	Square click
A0	Metronome click	Metronome click	Metronome click	Metronome click	Metronome click
A#0	Metronome bell	Metronome bell	Metronome bell	Metronome bell	Metronome bell
B0	Kick drum 2	Kick drum 2	Kick drum 2	Kick drum 2	Kick drum 2
C1	Kick drum 1	Kick drum 1	Mondo kick	Elec BD	Analog bass drum
C#1	Side stick	Side stick	Side stick	Side stick	Analog rim shot
D1	Snare drum 1	Snare drum 1	Gated SD	Elec SD	Analog snare drum
D#1	Hand clap	Hand clap	Hand clap	Hand clap	Hand clap
E1	Snare drum 2	Snare drum 2	Snare drum 2	Gated SD	Snare drum 2
F1	Low tom 2	Room lo tom 2	Room lo tom 2	Elec lo tom 2	Analog low tom 2
F#1	Closed hi-hat	Closed hi-hat	Closed hi-hat	Closed hi-hat	Analog closed hi-hat
G1	Low tom 1	Room lo tom 1	Room lo tom 1	Elec lo tom 1	Analog low tom 1
G#1	Pedal hi-hat	Pedal hi-hat	Pedal hi-hat	Pedal hi-hat	Analog closed hi-hat
A1	Mid tom 2	Room mid tom 2	Room mid tom 2	Elec mid tom 2	Analog mid tom 2
A#1	Open hi-hat	Open hi-hat	Open hi-hat	Open hi-hat	Analog closed hi-hat
B1	Mid tom 1	Room mid tom 1	Room mid tom 1	Elec mid tom 1	Analog mid tom 1
C2	High tom 2	Room hi tom 2	Room hi tom 2	Elec hi tom 2	Analog high tom 2
C#2	Crash cymbal	Crash cymbal	Crash cymbal	Crash cymbal	Analog cymbal
D2	High tom 1	Room hi tom 1	Room hi tom 1	Elec hi tom 1	Analog high tom 1
D#2	Ride cymbal	Ride cymbal	Ride cymbal	Ride cymbal	Ride cymbal
E2	Chinese cymbal	Chinese cymbal	Chinese cymbal	Reverse cymbal	Reverse cymbal
F2	Ride bell	Ride bell	Ride bell	Ride bell	Ride bell
F#2	Tambourine	Tambourine	Tambourine	Tambourine	Tambourine
G2	Splash cymbal	Splash cymbal	Splash cymbal	Splash cymbal	Splash cymbal

G#2	Cowbell	Cowbell	Cowbell	Cowbell	Analog cowbell
A2	Crash cymbal 2	Crash cymbal 2	Crash cymbal 2	Crash cymbal 2	Crash cymbal 2
A#2	Vibra-slap	Vibra-slap	Vibra-slap	Vibra-slap	Vibra-slap
B2	Ride cymbal 2	Ride cymbal 2	Ride cymbal 2	Ride cymbal 2	Ride cymbal 2
C3	High bongo	High bongo	High bongo	High bongo	High bongo
C#3	Low bongo	Low bongo	Low bongo	Low bongo	Low bongo
D3	Mute hi conga	Mute hi conga	Mute hi conga	Mute hi conga	Analog high conga
D#3	Open hi conga	Open hi conga	Open hi conga	Open hi conga	Analog mid conga
E3	Low conga	Low conga	Low conga	Low conga	Analog low conga
F3	High timbale	High timbale	High timbale	High timbale	High timbale
F#3	Low timbale	Low timbale	Low timbale	Low timbale	Low timbale
G3	High agogo	High agogo	High agogo	High agogo	High agogo
G#3	Low agogo	Low agogo	Low agogo	Low agogo	Low agogo
A3	Cabasa	Cabasa	Cabasa	Cabasa	Cabasa
A#3	Maracas	Maracas	Maracas	Maracas	Analog maracas
B3	Short hi whistle	Short hi whistle	Short hi whistle	Short hi whistle	Short hi whistle
C4	Long lo whistle	Long lo whistle	Long lo whistle	Long lo whistle	Long lo whistle
C#4	Short guiro	Short guiro	Short guiro	Short guiro	Short guiro
D4	Long guiro	Long guiro	Long guiro	Long guiro	Long guiro
D#4	Claves	Claves	Claves	Claves	Analog claves
E4	High woodblock	High woodblock	High woodblock	High woodblock	High woodblock
F4	Low woodblock	Low woodblock	Low woodblock	Low woodblock	Low woodblock
F#4	Mute cuica	Mute cuica	Mute cuica	Mute cuica	Mute cuica
G4	Open cuica	Open cuica	Open cuica	Open cuica	Open cuica
G#4	Mute triangle	Mute triangle	Mute triangle	Mute triangle	Mute triangle
A4	Open triangle	Open triangle	Open triangle	Open triangle	Open triangle
A#4	Shaker	Shaker	Shaker	Shaker	Shaker
B4	Jingle bell	Jingle bell	Jingle bell	Jingle bell	Jingle bell
C5	Belltree	Belltree	Belltree	Belltree	Belltree
C#5	Castanets	Castanets	Castanets	Castanets	Castanets
D5	Mute surdo	Mute surdo	Mute surdo	Mute surdo	Mute surdo
D#5	Open surdo	Open surdo	Open surdo	Open surdo	Open surdo

Table A1.5. *List of GM2 percussions for the first five kits*

Note (piano)	Jazz set	Brush set	Orchestra set	SFX set
D#0	High Q	High Q	Closed hi-hat	–
E0	Slap	Slap	Pedal hi-hat	–
F0	Scratch push	Scratch push	Open hi-hat	–
F#0	Scratch pull	Scratch pull	Ride cymbal	–
G0	Sticks	Sticks	Sticks	–
G#0	Square click	Square click	Square click	–
A0	Metronome click	Metronome click	Metronome click	–
A#0	Metronome bell	Metronome bell	Metronome bell	–
B0	Jazz BD 2	Jazz BD 2	Concert BD 2	–
C1	Jazz BD 1	Jazz BD 1	Concert BD 1	–
C#1	Side stick	Side stick	Side stick	–
D1	Snare drum 1	Brush tap	Concert SD	–
D#1	Hand clap	Brush slap	Castanets	High Q
E1	Snare drum 2	Brush swirl	Concert SD	Slap
F1	Low tom 2	Low tom 2	Tympani F	Scratch push
F#1	Closed hi-hat	Closed hi-hat	Tympani F#	Scratch pull
G1	Low tom 1	Low tom 1	Tympani G	Sticks
G#1	Pedal hi-hat	Pedal hi-hat	Tympani G#	Square click
A1	Mid tom 2	Mid tom 2	Tympani A	Metronome click
A#1	Open hi-hat	Open hi-hat	Tympani A#	Metronome bell
B1	Mid tom 1	Mid tom 1	Tympani B	Guitar fret noise
C2	High tom 2	High tom 2	Tympani C	Guitar cut noise up
C#2	Crash cymbal	Crash cymbal	Tympani C#	Guitar cut noise down
D2	High tom 1	High tom 1	Tympani D	Double bass string slap
D#2	Ride cymbal	Ride cymbal	Tympani D#	Flute key click
E2	Chinese cymbal	Chinese cymbal	Tympani E	Laughing
F2	Ride bell	Ride bell	Tympani F	Screaming
F#2	Tambourine	Tambourine	Tambourine	Punch
G2	Splash cymbal	Splash cymbal	Splash cymbal	Heartbeat
G#2	Cowbell	Cowbell	Cowbell	Footsteps 1

A2	Crash cymbal 2	Crash cymbal 2	Concert cymbal 2	Footsteps 2
A#2	Vibra-slap	Vibra-slap	Vibra-slap	Applause
B2	Ride cymbal 2	Ride cymbal 2	Concert cymbal 1	Door creaking
C3	High bongo	High bongo	High bongo	Door closing
C#3	Low bongo	Low bongo	Low bongo	Scratch
D3	Mute hi conga	Mute hi conga	Mute hi conga	Wind chimes
D#3	Open hi conga	Open hi conga	Open hi conga	Car engine
E3	Low conga	Low conga	Low conga	Car brakes
F3	High timbale	High timbale	High timbale	Car passing
F#3	Low timbale	Low timbale	Low timbale	Car crash
G3	High agogo	High agogo	High agogo	Siren
G#3	Low agogo	Low agogo	Low agogo	Train
A3	Cabasa	Cabasa	Cabasa	Jet plane
A#3	Maracas	Maracas	Maracas	Helicopter
B3	Short hi whistle	Short hi whistle	Short hi whistle	Starship
C4	Long lo whistle	Long lo whistle	Long lo whistle	Gun shot
C#4	Short guiro	Short guiro	Short guiro	Machine gun
D4	Long guiro	Long guiro	Long guiro	Laser gun
D#4	Claves	Claves	Claves	Explosion
E4	High woodblock	High woodblock	High woodblock	Dog bark
F4	Low woodblock	Low woodblock	Low woodblock	Horse gallop
F#4	Mute cuica	Mute cuica	Mute cuica	Birds tweet
G4	Open cuica	Open cuica	Open cuica	Rain
G#4	Mute triangle	Mute triangle	Mute triangle	Thunder
A4	Open triangle	Open triangle	Open triangle	Wind
A#4	Shaker	Shaker	Shaker	Seashore
B4	Jingle bell	Jingle bell	Jingle bell	Stream
C5	Belltree	Belltree	Belltree	Bubble
C#5	Castanets	Castanets	Castanets	–
D5	Mute surdo	Mute surdo	Mute surdo	–
D#5	Open surdo	Open surdo	Open surdo	–

Table A1.6. *List of GM2 percussions for the last four kits*

Appendix 2

MIDI Box, Merger and Patcher

This appendix brings together MIDI boxes, MIDI mergers and MIDI patchers, very practical little MIDI boxes for responding to unconventional MIDI connection situations.

If you want to directly connect several devices (devices 2, 3 and 4) from a single output (device 1), you can use a MIDI Thru Box as shown in Figure A2.1.

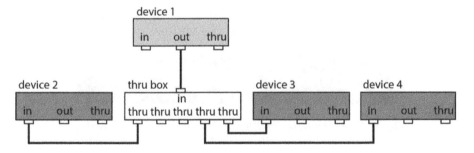

Figure A2.1. *Connections via a MIDI Thru Box*

The advantage of this device is that it eliminates chaining, which could degrade the signal by passing through many devices' IN and THRU sockets. With a MIDI Thru Box, the link is direct to each device.

Another option is to mix several MIDI inputs using a MIDI merger box, as shown in Figure A2.2.

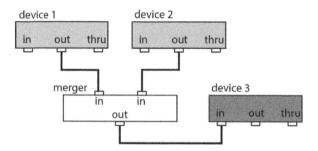

Figure A2.2. *Connections via a MIDI merger*

Another possible use is, for example, to control the same expander (hardware 3) with several master keyboards (hardware 1 and 2).

Much more sophisticated equipment includes MIDI patchers, MIDI patchbays or MIDI routers, whose features are very advanced; they can filter MIDI data, merge several inputs or several outputs and route signals to predefined outputs while storing the various settings and assignments in several memories that can be accessed directly later.

We can cite a few models such as the Roland A-880, the Digital Music Corp MX-8, the 360 Systems MIDI patcher or the Akai ME-30P.

Figure A2.3. *The Roland MIDI A-880 router. For a color version of this figure, see www.iste.co.uk/reveillac/synthesizers1.zip*

With the arrival of virtual instruments and effects on microcomputers, these routers have become rarer today, except for musicians who still have significant hardware configurations.

Glossary

Acoustic impedance: resistance of the medium (gas, liquid and solid) to the passage of an acoustic wave. Its unit is Pascal Second Per Meter (Pa s/m), also called a rayl.

ADSR (attack, decay, sustain, release): a set of parameters that form the envelope of a sound, that is, how it develops over time.

Aliasing: a phenomenon that introduces undesirable frequencies into a sampled signal or a signal modulating a carrier frequency. They appear when the carrier or sampling frequency is less than twice the maximum frequency of the signal.

Aliasing frequency: see aliasing.

Aperiodic signal: a signal whose form does not repeat itself, that is, that does not have a period.

Audio interface: sometimes called a sound card, it is a peripheral computer device that allows microphones, instruments and monitoring systems to be connected to a computer via a USB, PCI, Firewire or Thunderbolt port. It provides analog-to-digital conversion.

Auto-oscillation: also called self-oscillation. It is the generation and maintenance of periodic motion by a source devoid of corresponding periodicity. The oscillator itself controls the phase with which the external power supply acts on it.

Bandwidth: the frequency range of a sound source. We can also speak of sound spectrum width. It corresponds to the spectral size, that is, the interval between the low and high frequencies used by the source.

Beat: periodic modulation of an audio signal, consisting of two superimposed audio signals of close frequencies.

Bel: a logarithmic measurement unit of the ratio between two powers of an audio signal.

Bender: used to change the sound toward the bass or the treble. It can be done manually (wheel, joystick, etc.) or automatically via an envelope generator.

Binaural: is based on a sound capture method adapted to the morphology of the human head. This recording technique guarantees a very natural three-dimensional sound reproduction with excellent spatialization.

Brownian motion: a disordered and random movement, described mathematically, of a large particle immersed in a fluid or a gas, which is subjected to no other interaction than with the small molecules of the surrounding fluid.

Bypass: a mode that allows an audio signal to pass through a processing device without affecting its sound characteristics.

Carrier frequency: also called a carrier wave, or simply a carrier. It is a waveform that is modulated by an input signal (modulating signal or modulating frequency) in order to modify it (generally to apply an effect to it).

Carrier signal: see carrier frequency.

Celerity: also called speed. The speed of sound is the speed of propagation of sound waves in all media, gaseous, liquid or solid.

Click: high-amplitude, short-duration sound at the beginning of a waveform. It is found in music, noise or speech. It contains many non-periodic and high amplitude components at high frequencies. They often occur at the creation of a tone.

Combfilter: widely used in signal processing to add a delayed version of the signal to itself, which causes constructive or destructive interference.

Cutoff frequency: the limit frequency of useful operation of an electronic circuit, particularly for filters. High and low cutoff frequencies define the bandwidth.

CV/Gate (control voltage/gate): an electronic process for transmitting control information, used by many synthesizers and some other equipment such as sequencers or drum machines.

DAW: digital audio workstation.

DCO (digitally controlled oscillator): an oscillator whose frequency is digitally controlled. It is not a voltage that controls its frequency and amplitude but numerical values. The oscillator itself remains analog.

Decibel (dB): unit of magnitude, used in acoustics and electronics, defined as 10 times the decimal logarithm of the ratio between two powers.

Diaphony: interference between two audio signals.

Distortion: unwanted changes to an audio signal.

DO (digital oscillator): its frequency and waveform are created via software. It imitates an analog oscillator while eliminating or keeping its defects (stability over time).

Doppler effect: the frequency shift of an acoustic wave, observed between the measurements on transmission and reception when the distance between the transmitter and the receiver varies over time.

DSP (digital signal processing): a dedicated and optimized microprocessor to execute signal processing applications (filtering, extraction, conversion, encoding, decoding, compression, etc.).

Envelope: see ADSR.

Eurorack: a modular synthesizer format originally specified in 1996 by Doepfer Musikelektronik. It has made it possible to standardize the size and connectors of musical modules.

Firewire: trade name given by Apple to a multiplexed serial interface, also known by Sony under the name of i.Link, linked to the standard IEEE 1394.

Glide: ensures the transition from one note to another gradually. Most synthesizers have it; the passing speed between notes can be adjusted.

Grain: sound samples with a duration generally between 10 and 100 ms. These are the building blocks of granular sound synthesis. We control their density, pitch, length and envelope.

Haas effect: also called the "precedence effect," increases the stereophonic field. It is a psychoacoustic effect that uses a time shift imperceptible by the ear resulting in a greater field width.

Harmonic: also called a harmonic partial. It is a component of a periodic sound, the frequency of which is a multiple of the fundamental frequency. It can be even or odd.

Hertz: Hz is the International System (SI) unit of frequency. It is equivalent to one oscillation per second (s^{-1} or $1/s$).

HRTF (head-related transfer function): a transfer function characterized by the transformations brought to a sound signal by a listener's body, mainly the head and the ears. This allows humans to locate a sound in space, in azimuth (horizontally) and elevation (vertically).

ILD (interaural level difference): the intensity and frequency distribution difference between the two ears. Our ears can detect differences in loudness between the left and right ears, related to the acoustic shadow created by our head, which allows us to define the position of a sound source with a frequency higher than 1,500 Hz.

Interference: the combination of two sound waves of the same frequency. The interferences are the cause of diffraction and beat phenomena. The two types of interference, constructive and destructive, depend on the phase difference between the two sound waves.

ITD (interaural time difference): represents the difference in the arrival time of a sound wave at each ear. It is an important index to estimate the position of a sound source in the horizontal plane. It is effective in locating a source emitting waves whose frequency is lower than 1,500 Hz.

Key tracking: allows us to adapt the filter's cutoff frequency to the frequency of the note played. Thus, this cutoff frequency will no longer be fixed in relation to the entire sound spectrum but will vary for each note to obtain a more homogeneous rendering.

Keyboard splitter: divides the range of notes on a keyboard into several zones to assign them different sounds.

LFO (Low-frequency oscillator): a type of oscillator dedicated to low frequencies; its function is to control rather than to generate sound, that is, it can modulate another parameter.

Loudness: a numerical value representing the sound volume as perceived by a human being.

MIDI (Musical Instrument Digital Interface): protocol, communication standard and file format (MIDIfile) created for electronic musical instruments, which officially appeared in 1983. Its connectors are based on three five-pin DIN connectors: input (MIDI IN), output (MIDI OUT) and redirection (MIDI THRU). Two cables are required to ensure two-way communication between the various elements of the musical chain.

Modulation frequency: the frequency of one signal that modulates another. Modulation can be applied in frequency or amplitude.

Optocoupler: also called photo-coupler. It is an electronic component capable of transmitting the signal from one electrical circuit to another, without there being any galvanic contact between them.

Overdrive: name of an audio effect intended to recreate the saturation of an amplifier (usually an electric guitar) pushed to its gain limits.

Passband: the range of frequencies for which the attenuation of an audio signal is less than a specified value.

Phase: an instantaneous situation of a magnitude that varies cyclically. Two sound signals have different phases when they are of the same frequency and amplitude but are not superimposed.

Pitch-shifter: hardware or software tool, part of the sound effects, that allows for the modification of the pitch of a sound signal while remaining at the same tempo (without modifying its duration).

Presence: a device that adds treble and sharpness to sound. Generally, it acts directly on the amp's power stage via a feedback system (a system of reinjection of part of the signal at the amp's input).

Resonance: the tendency of an acoustic system to absorb more energy when the frequency of its oscillations reaches its natural frequency of vibration (resonant frequency).

Ring modulator: its principle is based on modulating the amplitude of a carrier wave by modulating a wave, which is often sinusoidal. If the modulating wave is audible (>20 Hz), new frequencies are obtained, replacing the carrier and modulating waves. These frequencies are the sum and the difference between the two previously mentioned waves.

Sample and hold: an electronic circuit consisting of a controlled switch and a capacitor. The purpose of a sample-and-hold is to sample an input signal at different times and then hold the last sample at a constant value.

Sampling: the digitization of a sound signal to create samples. It is carried out using a sampler.

Sound design: mastery and technical design of a soundtrack, within a framework defined upstream. The job of a sound designer is to develop a sound and artistic landscape that includes many elements, such as sound effects, music and artificial sounds.

Tessitura: set or scale of notes that the voice can emit without difficulty with the same sound volume and correct timbre.

Thunderbolt: connection and computer interface format designed by Intel, whose work began in 2007. The Mini Display Port connector, already present on Apple computers, was chosen as the standard interface for Thunderbolt. The latest versions (3 and 4) have a speed of 40 Gbit/s.

Timbre: sound parameters that characterize and identify an instrument or a voice.

Tonotopy: the spatial arrangement of the site where sounds of different frequencies are processed in the brain. Tonotopy in the auditory system begins in the cochlea, the small snail-shaped structure located in the inner ear, which sends information about the sound signal received by the ear to the brain.

Tremolo: periodic variation in the volume (and/or pitch) of a sound signal, usually produced by an LFO.

Universal serial bus (USB): the standard that defines a serial transmission linking several peripherals to a microcomputer or any other type of compatible device. It was designed in the mid-1990s and replaced many communication ports such as the parallel port, serial port and SCSI port. It is linked to a specific connection and has been able to evolve over time by increasing its speed, which can theoretically reach 10 Gbits/s in its latest version, USB 4, dating from 2017.

Vacuum tube: also called a "lamp" or "electronic tube". An active electronic component generally dedicated to the amplification of a signal. It was replaced later by semiconductors (transistor, diode, etc.). It was invented by Lee De Forest in 1907 (triode).

VCF (voltage-controlled filter): a module of an analog synthesizer including the filter part. This circuit is controlled by a voltage generated from a keyboard, sequencer or MIDI CV/Gate data.

VCO (voltage-controlled oscillator): a basic element of synthesis. It creates a waveform whose frequency and pitch are controlled by a voltage usually produced by a keyboard, sequencer or MIDI CV/Gate data.

Vibrato: periodic variation of the frequency of a sound signal generally produced by an LFO.

Waveshaping: an audio synthesis and transformation technique that transforms simple sounds into complex sounds. A pure sound, like a sine wave, can be transformed into a harmonically richer sound by changing its shape.

References

Anderton, C. (1986). *MIDI for Musicians*. The Music Sales Group, London.

Bianchi, F., Cipriani A., Giri, M. (2021). *Pure Data: Electronic Music and Sound Design – Theory and Practice – Volume 1*. ConTempoNet, Rome.

Braut, C. (1994). *Norme MIDI, Tome 1*. Sybex, Paris.

Braut, C. (1995). *Norme MIDI, Tome 2*. Sybex, Paris.

Chung, B. (2013). *Multiumedia Programming with Pure Data*. Packt Publishing, Birmingham.

Cohen-Levinas, D. (1999). *La synthèse sonore*. Éditions du Centre Pompidou, Paris.

Collins, N., Schedel, M., Wilson, S. (2013). *Electronic Music*. Cambridge University Press, Cambridge.

De Wilde, L. (2016). *Les fous du son*. Grasset, Paris.

Friedman, D. (1985). *Complete Guide to Synthesizers, Sequencers, and Drum Machines*. Music Sales Corp, California.

Grosse, D. (2021a). *Modular Synthesizer Mastery – Volume 1*. 20 Objects Publishing, Northfield.

Grosse, D. (2021b). *Modular Synthesizer Mastery – Volume 2*. 20 Objects Publishing, Northfield.

Grosse, D. (2021c). *Modular Synthesizer Mastery – Volume 3*. 20 Objects Publishing, Northfield.

Hillerson, T. (2014). *Programming Sound with Pure Data*. O'Reilly, Massachusetts.

Kendall, G. and Puddie Rodgers, C.A. (1981). The simulation of three-dimensional localization cues for headphone listening. *Proceedings of the International Computer Music Conference*, Denton, Texas.

Keyboard Magazine (1985). *Synthesizers and Computers*. Hal Leonard, Winona.

Lechner, P. (2014). *Multimedia Programming Using Max/MSP and TouchDesigner*. Packt Publishing, Birmingham.

Lehmann, H. (2017). *La Révolution digitale dans la musique*. Allia, Paris.

Manning, P. (2013). *Electronic and Computer Music*. OUP USA, Oxford.

Manzo, V.J. (2016). *Max/MSP/Jitter for Music: A Practical Guide to Developing Interactive Music Systems for Education and More*. OUP USA, Oxford.

Mercier, D. (ed.) (2002). *Le livre des techniques du son*. Dunod, Paris.

Meyer, C. and Brooks, E. (1994). MIDI time code, detailed specification. *Music Technology*, 1994 issue.

Pierce, J.R. (1984). *Le son musical*. Belin, Paris.

Pinch, T. and Trocco, F. (2002). *Analog Days: The Invention and Impact of the Moog Synthesizer*. Harvard University Press, Cambridge, Massachusetts.

Quinet, J.J. (1987). *Les cahiers de l'ACME. Le système MIDI*. Jean Jacques Quinet Éditeur, France.

Rausch, K. (2021). *The DX Story: FM Synthesis – The Magic Formula for the Sound of 80s*. NeoPubli GmbH, Berlin.

Réveillac, J.M. (2017). *Musical Sound Effects: Analog and Digital Sound Processing*. ISTE Ltd, London, and John Wiley & Sons, New York.

Roads, C. (2015). *Composing Electronic Music: A New Aesthetic*. OUP USA, Oxford.

Roads, C. (2016). *L'audionumérique – Musique et informatique*, 3rd edition. Dunod, Paris.

Rothstein, J. (1995). *MIDI: A Comprehensive Introduction – Volume 7*. A-R Editions, Wisconsin.

Taylor, G. (2018). *Step by Step: Adventures in Sequencing with Max/MSP*. Cycling'74, California.

Vail, M. (2014). *The Synthesizer: A Comprehensive Guide to Understanding, Programming, Playing, and Recording the Ultimate Electronic Music Instrument*. OUP USA, Oxford.

Vail, M. (2000). *Vintage Synthesizers*. Backbeat Books, Connecticut.

Internet links

Internet links are inherently volatile. They may evolve over time into other addresses, or even disappear. All were valid when writing this book; if some of them no longer work, a little research on Google or with your favorite search engine will help you find them.

Software manufacturers and publishers

Ableton: www.ableton.com

Akai: www.akaipro.com

Alesis: www.alesis.com

Arturia: www.arturia.com

Behringer: www.behringer.com

CTRLR: www.ctrlr.org

Cycling'74: www.cycling74.com

Dave Smith: www.davesmithinstruments.com

Doepfer: www.doepfer.de

E-mu: www.emu.com

Gforce Software: www.gforcesoftware.com

IK Multimedia: www.ikmultimedia.com

Korg: www.korg.com

Moog: www.moogmusic.com

MOTU: www.motu.com

Native Instruments: www.native-instruments.com

Novation: www.novationmusic.com

Synapse Audio Software: www.synapse-audio.com

U-He: www.u-he.com

Universal Audio: www.uaudio.com

VCV Rack: www.vcvrack.com

Yamaha: www.yamaha.com

Vintage electronic instruments

Synthmuseum, the website for old synthesizers: www.synthmuseum.com

Synthtopia, the museum for old synthesizers: www.synthtopia.com

The 14 most important synthesizers in electronic music: www.factmag.com/2016/09/15/14-most-important-synths

Vintage synthesizers: www.vintagesynth.com

Technical documentation

A website dedicated to the technical documentation of synthesizers: www.synthxl.com

A website dedicated, among other things, to the technical documentation of synthesizers: www.synthfool.com

General sites

A vintage synthesizers website: www.vintagesynth.com

Attack magazine, a music and digital audio magazine: www.attackmagazine.com

Audiofanzine, a website dedicated to music audio equipment: www.audiofanzine.com

Canford, a seller of audio and video products (in French): www.canford.fr

Harmony Central, one of the best websites dedicated to music audio: www.harmonycentral.com

Keyboard magazine, a music magazine: www.keyboardmag.com

KR Home studio, the music creation magazine (in French): www.kr-homestudio.fr

Music equipment: www.musicradar.com

Music equipment and more: www.pmtonline.co.uk

Music, musicians and instruments: www.factmag.com

Music Store Professional, seller of equipment: www.musicstore.de/en_OE/EUR

MusicTech, a website for sound engineers and for music production: www.musictech.net

Musiker Board, a music audio website (in German): www.musiker-board.de

Pro Audio Review magazine: www.prosoundnetwork.com/article.aspx?articleid=39995

ProSound: www.prosoundnetwork.com

SoundClick, a music audio website: www.soundclick.com

Tape Op magazine, a music magazine: www.tapeop.com

Thomann, a seller of online music products: https://www.thomann.de/gb/index.html

Training and resources for sound lovers: www.deveniringeson.com

Woodbrass, an online music seller: www.woodbrass.com

Zikinf, a general website for music and audio equipment: www.zikinf.com

Interfaces and communication

Digital audio connections (in French): www.mamosa.org/jenfi.home/debuter/connexions numeriques.php

Digital interfaces: www.soundonsound.com/techniques/digital-interfacing

Max/MSP

Ableton and Max for Live website: www.ableton.com/en/live/max-for-live

Cycling'74 publisher website: www.cycling74.com

Max objects database: www.maxobjects.com

Max/MSP course, introduction to Max/MSP: www.instructables.com/Intro-to-MaxMSP

Max/MSP course, vision and sound with Max: www.goldbergs.com/max/parsons

Max/MSP forum: www.cycling74.com/forums/page/1

Patcher for Max/MSP, Max/MSP Codelab: www.codelab.fr/max-msp

vmpk, virtual MIDI keyboard: www.vmpk.sourceforge.io/#Introduction

MIDI

An introduction to the MIDI standard: www.midi.org/articles-old/an-intro-to-midi

An introduction to the MIDI standard (in French): www.cri.ensmp.fr/~pj/music_slides.pdf

Differences between GM, GS and XG: www.cybermidi.com/helpdesk/knowledgebase.php?article=47

Exploring the GM standard: www.harfesoft.de/aixphysik/sound/midi/pages/genmidi.html

General MIDI (in French): www.daffyduke.lautre.net/zik/midi_10.html

MIDI 2.0: www.midi.org/midi-articles/details-about-midi-2-0-midi-ci-profiles-and-property-exchange

MIDI forum: www.midi.org/forum/830-midi-octave-and-note-numbering-standard

The GM standard: www.midi.org/specifications-old/item/gm-level-1-sound-set

The MIDI standard (in French): www.ntemusique.free.fr/musique/MIDI/MIDI.pdf

The MIDI standard (in French): www.ogloton.free.fr/midi/presentation.html

The MIDI standard (in French): www.sonelec-musique.com/electronique_theorie_midi_norme.html

The MIDI standard and its files (in French): www.jchr.be/linux/midi-format.htm

The MIDI XG format (in French): www.ppretot.free.fr/whatxgf.htm

Pure Data

Noisy box, Pure Data Abstractions: https://noisybox.net/pd

pad.scope, an oscilloscope for Pure Data: www.patchstorage.com/pad-library

Pure Data: www.puredata.info

Pure Data Forum: www.forum.pdpatchrepo.info

Pure Data GitHub: www.github.com/pure-data/pure-data

Resources for Pure Data: www.mortmain.com/pd.html

Resources for Pure Data, Floss Manuals Pure Data: www.fr.flossmanuals.net/puredata/introduction

Virtual keyboard

A piano keyboard in Pure Data from your AZERTY keyboard: www.puredata.info/Members/anfex/teclado/view?searchterm=keyboard

A virtual MIDI keyboard in Pure Data: www.puredata.info/Members/fr4nck/fr4-azerty.pd/view?searchterm=keyboard

Keybin, a MIDI keyboard in Pure Data from your microcomputer keyboard: www.puredata.info/Members/sokratesla/abstractions/keybin.pd/view?searchterm=keyboard

vmpk, the virtual MIDI keyboard for MacOS, Microsoft Windows and Linux: www.vmpk.sourceforge.io/#Introduction

Reaktor blocks

Modules for Reaktor – Toybox: www.toyboxaudio.com

Native Instruments Blocks Wired: www.audiopluginsforfree.com/native-instruments-blocks-wired/

Tutorial – Getting started with Reaktor blocks: www.ask.video/article/audio-software/getting-started-with-reaktor-blocks

Tutorial – How to build a synth in Reaktor: www.adsrsounds.com/reaktor-tutorials/how-to-build-a-synth-in-reaktor-a-beginners-guide

Tutorial – How to build your first Reaktor synth: www.musicradar.com/news/how-to-build-a-reaktor-synth

Video tutorial – Reaktor know-how – Blocks: www.groove3.com/tutorials/REAKTOR-Know-How-Blocks

Working with Reaktor Blocks – Processing audio with Reaktor Blocks: www.macprovideo.com/article/audio-software/processing-audio-with-ni-reaktor-blocks

Software synthesizers

Arturia ARP 2600 V: www.arturia.com/products/analog-classics/arp2600v

Best virtual Minimoogs: www.careersinmusic.com/moog-vst

Best VST/AU Minimoog emulation: www.musicradar.com/news/tech/6-of-the-best-vst-au-minimoog-emulation-plugins-622621

Cherry Audio CA2600: www.cherryaudio.com/products/ca2600

CTRLR, control your MIDI life –MIDI publisher: www.ctrlr.org

MiniMogueVA: www.plugins4free.com/plugin/405

MiniMoog, Arturia Mini V: www.arturia.com/products/analog-classics/mini-v/overview

Native Instruments Reaktor Blocks: www.native-instruments.com/fr/products/komplete/synths/reaktor-6/blocks

Sonivox TimeWarp 2600: www.sonivoxmi.com/virtual-instruments/time-warp-2600.html

VCV Rack: www.vcvrack.com

VCV Rack – Rack Github: www.github.com/VCVRack/Rack

VST emulation of famous synths: www.productionmusiclive.com/blogs/news/top-5-free-emulations-of-famous-synths

Sound synthesis

A granular synthesis resource: www.granularsynthesis.com/index.php

A slideshow on reverb: www.cs.wellesley.edu/~cs203/lecture_materials/reverb/reverb.pdf

ADSRSounds – Theory behind FM synthesis: www.adsrsounds.com/fm8-tutorials/theory-behind-fm-synthesis/

An introduction to synth programming for electronic music producers: www.renegadeproducer.com/audio-synthesis.html

AudioKeys – Sound synthesis and synthesizers forum (in French): www.audiokeys.net/forum

AudioShapers – Noise, types and colors: www.audioshapers.com/blog/noise-types-and-colors.html

Different types of noise: www.emastered.com/fr/blog/different-types-of-noise

FM synthesis (in French): www.inrp.fr/JIPSP/phymus/m_techni/synthfm/ac_syfm.htm

Freeverb: www.dsprelated.com/freebooks/pasp/Freeverb.html

Granular synthesis: www.sfu.ca/~truax/gran.html

History of sound synthesis (in French): www.frank-lovisolo.fr/WordPress/synthese-sonore-histoire-raccourci

Izotope – Basics of FM synthesis: www.izotope.com/en/learn/simple-fm-synthesis-sine-waves-and-processors.html

Izotope – The basics of granular synthesis: www.izotope.com/en/learn/the-basics-of-granular-synthesis.html

Music-tech – Learning the base of FM synthesis: www.musictech.com/guides/essential-guide/how-fm-synthesis-works/

Musical signal processing with LabVIEW: www.cnx.org/contents/4ODlu9s3@1.6:H3hx-Qm8@3/Schroeder-Reverberator

Natural sounding artificial reverberation: www.docplayer.net/58419605-Natural-sounding-artificial-reverberation.html

Noise and colors (in French): www.easyzic.com/dossiers/les-bruits,h391.html

Principles of sound synthesis (in French): www.easyzic.com/dossiers/principes-de-la-synthese,h392.html

Processing of audio signals: www.pact.wp.imt.fr/files/2011/09/EffetsetReverbGrichard.pdf

Schroeder reverberators: www.ccrma.stanford.edu/~jos/pasp/Schroeder_Reverberators.html

Schroeder reverberators: www.dsprelated.com/freebooks/pasp/Schroeder_Reverberators.html

Sound on Sound – Granular synthesis: www.soundonsound.com/techniques/granular-synthesis

Sound synthesis (in French): www.traitement-signal.com/synthese_sonore.php

Sound synthesis part 1: www.soundonsound.com/techniques/sound-synthesis-part-1

Sound synthesis part 2: www.soundonsound.com/techniques/sound-synthesis-part-2

SoundFly – An introduction to FM synthesis: www.flypaper.soundfly.com/produce/an-introduction-to-fm-synthesis/

Synthesizers (in French): www.frank-lovisolo.fr/WordPress/synthese-sonore-histoire-raccourci

The basics of sound synthesis: www.theproaudiofiles.com/sound-synthesis-basics

The beginning of sound synthesis: www.digitalsoundandmusic.com/6-1-1-the-beginnings-of-sound-synthesis/

Understanding synthesizers: www.splice.com/blog/understanding-synthesizers/

Various video tutorials

Elephorm: www.elephorm.com/audio-mao.html

Learning modular – Courses on modular synthesizers: www.learningmodular.com

Tutorom (in French): www.tutorom.fr/categories-de-tutoriels/fr/audio

Virtual production school (VPS) (in French): www.tutoriels-mao.com/les-tutoriels/mix-et-master-de-a-%C3%A0-z-avec-des-plugins-gratuits-detail

Index

A, B

Ableton, 159, 170
absorption, 52–54
   coefficient, 53, 54
acoustic impedance, 51, 239
aftertouch, 132, 141, 159, 189, 200, 201
algorithms, 71, 80, 81, 94
aliasing, 89, 239
all-pass, 111
amplitude, 7, 9, 11, 22, 24, 25, 37, 46, 47, 74, 79, 88, 90, 96–98, 104, 112, 117, 121–123, 240, 241, 243
aperiodic, 22, 27, 58, 239
apex, 12, 14
arpeggiator, 141, 150, 157–160, 162, 164, 166
   -sequencer, 141
asymmetrical, 29, 30, 35
attack, 40, 41, 43, 85, 94, 114–116, 139, 151, 158, 166, 192, 226, 239
attenuation, 14, 35, 53, 105–109, 158, 185, 243
auditory nerve, 13
auricle, 11, 18, 20
band-pass, 105, 108, 111, 153, 163
bandwidth, 75, 89, 108, 109, 111, 239, 240

beat, 57, 58, 77, 210, 240, 242
Behringer, 118, 130, 131, 137, 138, 142, 143, 145, 147, 150, 152–155, 173, 214
bias function, 154
binaural, 15, 20, 240
blocks, 80, 145, 151, 155, 160, 166–168, 203–206, 209, 211, 222, 241
Bluetooth, 135
bone conduction, 14
breath controllers, 132

C, D

canal, 11, 18, 195, 212, 222
canvas, 172, 173
carrier, 4, 78–83, 85, 86, 96, 97, 120–122, 239, 240, 243
   signal, 78, 240
channel, 182, 186–189, 191–193, 195–197, 199–202, 204–206, 211, 212, 219, 221, 232
chorus, 35, 105, 124, 128, 129, 150, 164, 166, 193, 202, 209, 212, 225
chunks, 203
ciliated cells, 13
clipping, 124

cochlea, 11, 13, 244
cochlear duct, 11, 13, 14
comb filter, 111
common messages, 190
concave, 50–52, 54
continuous, 23, 27, 60, 71, 88, 116, 125, 153, 191
control change, 189
converter, 89, 213
convex, 50, 54
daisy chain, 184
damping, 41
data bytes, 187, 188, 194, 203, 205
decay, 41, 53, 85, 114, 115, 127, 139, 151, 166, 192, 239
delay, 21, 116, 124, 127–129, 153–155, 158, 164, 166, 182, 192, 205
detune, 132, 193
difference, 16–18, 21, 31, 49, 57, 58, 75, 96, 98, 107, 120, 121, 242, 243
diffraction, 18, 48, 49, 242
diffusion, 65, 90
digital oscillator (DO), 1, 2, 43, 44, 48, 60, 77, 95, 99, 126, 130, 137, 138, 167, 175, 176, 185, 188, 195, 210, 211
dispersion, 44, 45
dithering, 61
duct, 11, 13, 14

E, F

ear, 1, 5, 6, 9–12, 14, 15, 18, 20, 21, 26, 40, 43, 46, 62, 75, 78, 117, 241, 242, 244
eardrum, 10, 11, 18
echo, 21, 41, 52, 127, 128, 229
elasticity, 4, 90, 91
endolymph, 13

envelope, 27, 41, 72–74, 77, 80, 83–85, 87, 88, 90, 94–97, 99–101, 112–118, 130, 131, 139, 145, 149, 154, 158, 164, 166, 176, 215, 239–241
  generator, 72, 74, 77, 80, 83, 101, 112–118, 145, 149, 166, 240
Eurorack, 150, 155, 169, 214, 241
Eustachian tube, 11
exciter, 90
expression pedal, 101, 133
feedback, 86, 110, 128, 225, 243
field, 2, 19, 53, 99, 137, 180, 216, 241
filter, 72–75, 77, 103, 105–113, 117, 126, 130, 133, 139–141, 144, 145, 151, 153–159, 162, 163, 166, 238, 242, 245
flanger, 105, 124, 128, 164, 166
frequency, 1, 5–7, 9, 14, 15, 18, 22, 24–32, 35, 37, 39, 46, 53, 55–62, 64, 66, 71–80, 82, 85, 88–90, 95, 96, 98, 99, 101–111, 113, 116, 121, 122, 126, 129, 130, 139, 154, 158, 159, 163, 164, 178, 190, 239–243, 245

G, H

gate, 102, 135, 145, 148, 150, 151, 162, 166, 175, 176, 180, 213–215, 240, 245
glide, 124, 125, 130, 149, 214, 241
grain, 94, 95, 241
group, 42, 44, 77
Hammond, 67, 72, 75, 224
harmonics, 25–27, 29–32, 35, 37, 39, 41, 44, 71, 74, 76, 77, 79, 86, 98, 99, 107, 108, 110, 111, 116, 122, 220, 225

hertz (Hz), 5, 7, 9, 11, 17, 26, 28, 54, 58, 59, 63, 75, 76, 78, 82, 85, 86, 97, 103, 104, 109, 121, 154, 164, 176, 178, 179, 196, 197, 200, 214, 242, 243
high-pass, 105, 108–110, 153, 163
hold, 101, 116, 118, 120, 145, 159, 166, 192, 202, 244
Huygens, 48

I, K, L

incus, 11
inlets, 170
intensity, 3–5, 9, 15, 17, 18, 26, 27, 30, 31, 35, 37, 39, 41, 46, 50, 57–61, 65, 74, 75, 82, 83, 90, 105, 110, 114, 115, 117, 132, 192, 242
key follow, 127
keyboard tracking, 124, 126
keytar, 132, 134
legato, 125, 130, 192
level, 4–6, 9, 15–17, 31, 45, 85, 86, 114, 115, 119, 120, 141, 153, 164, 173, 193, 196, 200, 201, 242
loudness, 4, 15, 27, 39, 62, 242
low-pass, 105–111, 153, 158, 163

M, N

malleus, 11
mass, 42, 43, 90, 91, 95
Mellotron, 68, 69
membrane, 10, 13, 42, 90, 91
MIDI, 183, 190, 196, 211
Minimoog, 72, 102, 115, 124, 137, 147–149, 151, 152, 160, 162, 173, 215
modulation, 35, 73, 78–80, 96–98, 101, 102, 121, 122, 128, 129, 132, 142, 148–150, 154, 157, 158, 160, 163, 164, 166, 187, 192, 200, 201, 240, 243

modulator, 78, 80–83, 86, 96, 157, 162, 163
modwheel, 132
monody, 129, 130
Moog, 65, 70, 72, 106, 125, 134, 138, 147, 148, 152, 160, 163, 176, 186
morphology, 18, 240
multipole, 112
multitimbrality, 131
newton, 4
noise, 2, 9, 22–24, 44, 58–63, 72, 77, 79, 94, 99, 114, 119, 139, 145, 150, 151, 155, 157, 159, 163, 166, 221, 230, 231, 235, 240
notch, 105, 110, 111
note-off, 84, 194, 195, 212
note-on, 84, 194, 195, 212
notes, 2, 103, 116, 119, 125, 126, 129, 130, 159, 164, 177, 179, 189, 193, 200, 201, 205, 210, 211, 222, 241, 242, 244
Novachord, 72

O, P

omni, 187–189, 193
ondioline, 67, 68, 215
operators, 80–82, 85, 86, 170
oscillators (*see also* self-oscillation), 64, 65, 68, 72–76, 89, 96, 98, 101–104, 129, 131, 138, 139, 144, 145, 150, 153, 155–157, 159, 162, 163, 166, 167, 176, 193
outlets, 170
oval window, 11
overdrive, 153, 155, 158, 243
paraphonic, 130, 131, 153, 155
pascal, 4, 239
patcher, 170, 171, 237, 238
patterns, 158, 204
percussions, 44, 222, 234, 236
perilymph, 11

periodic, 7, 22, 24, 25, 27, 58, 74, 88, 239, 240, 242, 244, 245
pharynx, 11
phase, 17, 31, 37, 40, 44, 46, 49, 51, 57, 58, 75, 76, 98, 99, 108, 111, 128, 164, 223, 239, 242, 243
   change, 51, 75
   distortion, 98
   inversion, 51
phaser, 105, 124, 128, 193
pitch, 3, 7, 26, 28, 29, 42, 44, 58, 65, 66, 77–80, 87, 95, 103, 108, 124, 127, 148, 158, 163, 175, 176, 189, 214, 221, 241, 243–245
point of audition, 44
poles, 106, 107, 158, 163
polyphony, 101, 103, 129–131, 154, 169, 189, 199, 202
portamento, 125, 126, 130, 142, 158, 160, 192, 201
pressure, 4, 5, 7, 9, 11, 15, 17, 18, 45, 50, 90, 102, 132, 133, 157, 162, 166, 189
program, 93, 182, 189, 195, 199, 208, 209, 219, 223, 232
propagation, 2, 44, 50, 55, 91, 127, 215, 240
Prophet 600 synthesizer, 181
psophometric, 7, 62
pulsar train, 99
pulsaret, 99
pulse wave, 35, 154

R, S

random walk, 60
range, 1, 7, 41, 42, 63, 66, 75, 99, 102, 103, 108–110, 124, 154, 157, 171, 176, 191, 193, 215, 239, 242, 243
Reaktor, 167, 168

reflection, 48, 50, 51, 54, 55, 127
reflex latency, 11
refraction, 48, 55
rejector, 105, 163
release, 40, 53, 85, 91, 94, 114–116, 139, 158, 162, 166, 192, 195, 239
resonance, 65, 85, 110, 111, 116, 154, 155, 158, 243
resonator, 72, 90
ring modulator, 120, 121, 139, 145, 243
samplers, 68, 69
sampling, 88–90, 99, 100, 119, 127, 239, 244
Schaeffer, 42, 43
selectivity, 109–111
self-oscillation, 111, 158, 239
sequence number, 190, 208
Shannon, 10, 89
single note, 129
slope, 40, 85, 108
sound(s)
   classification of, 43
   curves, 15
   design, 2, 215, 244
   speed of, 4, 7, 9, 17, 52, 56, 240
spatial listening, 15
spectral
   analysis, 2, 25
   folding, 89
spectrum, 9, 14, 18, 22, 25–27, 31, 36, 39, 42, 44, 59–63, 77, 86, 87, 89, 96, 97, 121, 239, 242
stapedius muscle, 11
stapes, 11
superposition principle, 47
sustain, 40, 41, 85, 94, 114–116, 120, 133, 139, 151, 166, 167, 192, 200, 239
symmetrical, 29, 31

Synket, 72, 73
synthesis, 1, 2, 35, 37, 41, 58, 61, 63, 66–75, 77–80, 82–85, 88, 90, 93–96, 98–100, 104, 106, 111, 116, 122, 124, 137, 138, 148, 162, 167, 173, 182, 199, 215, 216, 241, 245
sysex event, 206
system exclusive, 190, 200, 206, 207

T, U

table, 3, 6, 26, 44, 53, 54, 82, 86, 87, 89, 90, 94, 99, 143, 145, 147, 151, 152, 159, 177–179, 187–191, 193–197, 199, 203–205, 207, 211, 213, 214, 219, 221, 222, 232, 234, 236
tempo map, 205
terminology, 129
theremin, 64, 65, 215
ticks, 204, 211
timbre, 2, 3, 9, 10, 18, 20, 39, 43, 75, 78–80, 83, 85, 105, 107, 109, 122, 131, 192, 199, 215, 244
tone, 58, 67, 85, 151, 153, 157, 178, 211, 225, 240
tonotopy, 14, 244

transients, 40, 113
transmutation, 112
tremolo, 40, 102, 105, 124, 193, 221, 226, 244
trigger, 113, 114, 116–119, 130, 131, 135, 176, 180
typology, 2, 9, 21, 22
unison, 132

V, W

velocity, 4, 114, 132, 150, 157–159, 162, 164, 166, 182, 189, 193, 195, 212, 213
vestibule, 11
vibrato, 40, 68, 78, 102, 105, 124, 129, 192, 200, 245
Victor (RCA Mark II model), 68
voice messages, 189, 201, 202
voltage trigger, 180
volts per octave, 176–178
Vox, 68, 69, 229
wavefolding, 124
wavelength, 7, 9, 48–50, 55, 56
waveshaping, 101, 122, 124, 245
wavewrapping, 124
wet/dry, 154

Other titles from

in

Waves

2022

GONTRAND Christian
Electromagnetism: Links to Special Relativity

ROYER Daniel, VALIER-BRASIER Tony
Elastic Waves in Solids 1: Propagation

VALIER-BRASIER Tony, ROYER Daniel
Elastic Waves in Solids 2: Radiation, Scattering, Generation

2021

DAHOO Pierre-Richard, LAKHLIFI Azzedine
Infrared Spectroscopy of Symmetric and Spherical Top Molecules for Space Observation 1
(Infrared Spectroscopy Set – Volume 3)
Infrared Spectroscopy of Symmetric and Spherical Top Molecules for Space Observation 2
(Infrared Spectroscopy Set – Volume 4)

LACAZE Pierre Camille, LACROIX Jean-Christophe
Nanotechnology and Nanomaterials for Energy

RÉVEILLAC Jean-Michel
Recording and Voice Processing 1: History and Generalities
Recording and Voice Processing 2: Working in the Studio

SAKHO Ibrahima
Nuclear Physics 1: Nuclear Deexcitations, Spontaneous Nuclear Reactions

2020

DANIELE Vito G., LOMBARDI Guido
Scattering and Diffraction by Wedges 1: The Wiener-Hopf Solution - Advanced Applications
(Waves and Scattering Set – Volume 1)
Scattering and Diffraction by Wedges 2: The Wiener-Hopf Solution - Advanced Applications
(Waves and Scattering Set – Volume 2)

SAKHO Ibrahima
Introduction to Quantum Mechanics 2: Wave-Corpuscle, Quantization & Schrödinger's Equation

2019

BERTRAND Pierre, DEL SARTO Daniele, GHIZZO Alain
The Vlasov Equation 1: History and General Properties

DAHOO Pierre-Richard, LAKHLIFI Azzedine
Infrared Spectroscopy of Triatomics for Space Observation
(Infrared Spectroscopy Set – Volume 2)

RÉVEILLAC Jean-Michel
Electronic Music Machines: The New Musical Instruments

ROMERO-GARCIA Vicente, HLADKY-HENNION Anne-Christine
Fundamentals and Applications of Acoustic Metamaterials: From Seismic to Radio Frequency
(Metamaterials Applied to Waves Set – Volume 1)

SAKHO Ibrahima
Introduction to Quantum Mechanics 1: Thermal Radiation and Experimental Facts Regarding the Quantization of Matter

2018

SAKHO Ibrahima
Screening Constant by Unit Nuclear Charge Method: Description and Application to the Photoionization of Atomic Systems

2017

DAHOO Pierre-Richard, LAKHLIFI Azzedine
Infrared Spectroscopy of Diatomics for Space Observation (Infrared Spectroscopy Set – Volume 1)

PARET Dominique, HUON Jean-Paul
Secure Connected Objects

PARET Dominque, SIBONY Serge
Musical Techniques: Frequencies and Harmony

RÉVEILLAC Jean-Michel
Analog and Digital Sound Processing

STAEBLER Patrick
Human Exposure to Electromagnetic Fields

2016

ANSELMET Fabien, MATTEI Pierre-Olivier
Acoustics, Aeroacoustics and Vibrations

BAUDRAND Henri, TITAOUINE Mohammed, RAVEU Nathalie
The Wave Concept in Electromagnetism and Circuits: Theory and Applications

PARET Dominique
Antennas Designs for NFC Devices

PARET Dominique
Design Constraints for NFC Devices

WIART Joe
Radio-Frequency Human Exposure Assessment

2015

PICART Pascal
New Techniques in Digital Holography

2014

APPRIOU Alain
Uncertainty Theories and Multisensor Data Fusion

JARRY Pierre, BENEAT Jacques N.
RF and Microwave Electromagnetism

LAHEURTE Jean-Marc
UHF RFID Technologies for Identification and Traceability

SAVAUX Vincent, LOUËT Yves
MMSE-based Algorithm for Joint Signal Detection, Channel and Noise Variance Estimation for OFDM Systems

THOMAS Jean-Hugh, YAAKOUBI Nourdin
New Sensors and Processing Chain

TING Michael
Molecular Imaging in Nano MRI

VALIÈRE Jean-Christophe
Acoustic Particle Velocity Measurements using Laser: Principles, Signal Processing and Applications

VANBÉSIEN Olivier, CENTENO Emmanuel
Dispersion Engineering for Integrated Nanophotonics

2013

BENMAMMAR Badr, AMRAOUI Asma
Radio Resource Allocation and Dynamic Spectrum Access

BOURLIER Christophe, PINEL Nicolas, KUBICKÉ Gildas
Method of Moments for 2D Scattering Problems: Basic Concepts and Applications

GOURE Jean-Pierre
Optics in Instruments: Applications in Biology and Medicine

LAZAROV Andon, KOSTADINOV Todor Pavlov
Bistatic SAR/GISAR/FISAR Theory Algorithms and Program Implementation

LHEURETTE Eric
Metamaterials and Wave Control

PINEL Nicolas, BOURLIER Christophe
Electromagnetic Wave Scattering from Random Rough Surfaces: Asymptotic Models

SHINOHARA Naoki
Wireless Power Transfer via Radiowaves

TERRE Michel, PISCHELLA Mylène, VIVIER Emmanuelle
Wireless Telecommunication Systems

2012

LALAUZE René
Chemical Sensors and Biosensors

LE MENN Marc
Instrumentation and Metrology in Oceanography

LI Jun-chang, PICART Pascal
Digital Holography

2011

BECHERRAWY Tamer
Mechanical and Electromagnetic Vibrations and Waves

BESNIER Philippe, DÉMOULIN Bernard
Electromagnetic Reverberation Chambers

GOURE Jean-Pierre
Optics in Instruments

GROUS Ammar
Applied Metrology for Manufacturing Engineering

LE CHEVALIER François, LESSELIER Dominique, STARAJ Robert
Non-standard Antennas

2010

BEGAUD Xavier
Ultra Wide Band Antennas

MARAGE Jean-Paul, MORI Yvon
Sonar and Underwater Acoustics

2009

BOUDRIOUA Azzedine
Photonic Waveguides

BRUNEAU Michel, POTEL Catherine
Materials and Acoustics Handbook

DE FORNEL Frédérique, FAVENNEC Pierre-Noël
Measurements using Optic and RF Waves

FRENCH COLLEGE OF METROLOGY
Transverse Disciplines in Metrology

2008

FILIPPI Paul J.T.
Vibrations and Acoustic Radiation of Thin Structures

LALAUZE René
Physical Chemistry of Solid-Gas Interfaces

2007

KUNDU Tribikram
Advanced Ultrasonic Methods for Material and Structure Inspection

PLACKO Dominique
Fundamentals of Instrumentation and Measurement

RIPKA Pavel, TIPEK Alois
Modern Sensors Handbook

2006

BALAGEAS Daniel *et al.*
Structural Health Monitoring

BOUCHET Olivier *et al.*
Free-Space Optics

BRUNEAU Michel, SCELO Thomas
Fundamentals of Acoustics

FRENCH COLLEGE OF METROLOGY
Metrology in Industry

GUILLAUME Philippe
Music and Acoustics

GUYADER Jean-Louis
Vibration in Continuous Media

Printed and bound by CPI Group (UK) Ltd, Croydon, CR0 4YY
26/03/2024

14476304-0001